English for Technical Purposes

Hans-Jürgen Bauer

Cornelsen & OXFORD

English for Technical Purposes wurde geplant und entwickelt von der Verlagsredaktion der Cornelsen & Oxford University Press GmbH & Co., Berlin.

Verfasser:	Hans-Jürgen Bauer, Darmstadt
Berater:	Dr Georg Aigner, Landshut
Redaktionelle Mitarbeit:	John Stevens, Fritz Preuß
Verlagsredaktion:	James Austin
Design:	Oxford Designers & Illustrators
Herstellung:	Axel Jens Penner
Bildredaktion:	Uta Hübner

Erhältlich sind auch: Textcassette (Best.-Nr. **25055**)
Lehrerhandbuch (Best.-Nr. **25047**)

1. Auflage ✔
5. 4. 3. 2. 1. Die letzten Ziffern bezeichnen
04 03 02 01 00 Zahl und Jahr des Druckes.

Alle Drucke dieser Auflage können, weil untereinander unverändert, im Unterricht nebeneinander verwendet werden.

Bestellnummer **25039**

ISBN 3-8109-2503-9

Druck & Weiterverarbeitung: Druckerei zu Altenburg

Vertrieb: Cornelsen Verlag, Berlin

Gedruckt auf chlorfrei gebleichtem Papier ohne Dioxinbelastung der Gewässer.

Vorwort

English for Technical Purposes ist ein völlig neu entwickeltes Lehrwerk für den berufsorientierten technischen Englischunterricht in der Aus- und Weiterbildung. Es basiert auf dem Niveau einer guten mittleren Reife bzw. eines erfolgreichen Abiturabschlusses und setzt einen Grundwortschatz von etwa 2000 Wörtern voraus. Das Lehrwerk führt systematisch zu den Kommunikationsfähigkeiten, die in weiten Bereichen der Lehre, der Industrie und des Handwerks benötigt werden. Es bereitet damit auch auf den Erwerb internationaler Sprachenzertifikate, wie u. a. das VHS/ICC-Zertifikat *English for Technical Purposes*, vor.

Aufbau

English for Technical Purposes gliedert sich in zwei Hauptteile: *Foundation Course* und *Modules*. Der *Foundation Course* umfasst sechs *Units*; parallel dazu stehen zwölf *Modules* zur Verfügung, die paarweise zusammengefasst sind.
Je eine *Unit* und ein *Module*-Paar, z. B. *Unit 1* und *Modules 1a/b*, sind einem der folgenden sechs Gebiete der Technik zugeordnet: *Manufacturing, Energy, Electricity, Architecture, Recycling* und *Telecommunications*.

In den sechs *Units* des *Foundation Course* liegt sowohl das sprachliche wie auch das technische Anspruchsniveau niedriger. Grundwortschatz und -grammatik werden wiederholt und reaktiviert und der neue Wortschatz wird in einem breiten, technisch relevanten Anwendungsspektrum vermittelt. Sprechanlässe von allgemein beruflicher Relevanz wie *Describing a company, Arranging a visit, Describing a product, Telephoning* usw. sind in eine wirklichkeitsnahe Rahmengeschichte eingebettet. Am Schluss jeder *Unit* befindet sich eine Übersicht mit interkulturellen Informationen aus verschiedenen Industrieländern, in denen Englisch häufig als Verkehrssprache benutzt wird. Thema ist jeweils das landesspezifische Verhalten in einem wichtigen sozialen Bereich des Berufslebens: *Welcoming a visitor to a company, Etiquette of meetings* usw.

In den *Modules* liegt das technische Anspruchsniveau in den genannten Fachgebieten (von *Manufacturing* bis *Telecommunications*) höher. Schwerpunkte sind hier der Umgang mit authentischen Texten, das Lösen technischer Probleme in der Fremdsprache und die Aneignung von Lerntechniken (vom Umgang mit einem Wörterbuch bis hin zum Prüfungstraining). Die *Units* und *Modules* folgen keiner strengen sprachlichen Progression und können deshalb unabhängig voneinander bearbeitet werden. Je nach Unterrichtssituation können sowohl Auswahl als auch Reihenfolge den individuellen Bedürfnissen und zeitlichen Voraussetzungen der Lernenden flexibel angepasst werden.
Im Anhang befinden sich die Transkriptionen der Hörübungen, die Rollenspielanweisungen, eine Übersicht über die sprachliche Umsetzung mathematischer Zeichen und Gleichungen sowie eine Umwandlungstabelle für Maßeinheiten.

Inhalt und Ziele

Alle wichtigen Fertigkeiten und Kenntnisse werden geschult: *Reading, Listening, Speaking, Writing, Grammar, Vocabulary, Culture, Social Skills* und *Study Skills*. Die Sprachfertigkeiten werden anhand einer breiten Vielfalt von Textsorten (authentisch und leicht adaptiert) trainiert. Es werden angeboten: Auszüge aus Firmenbroschüren, Lehrbüchern, populärwissenschaftlichen Zeitschriftenartikeln und Betriebsanleitungen sowie Geschäftsbriefe, -faxe und E-Mails, Lebenslauf, Vortrag, Präsentation, Telefongespräche usw. Diese Textsorten bieten die Plattform für das Üben des freien Sprechens u. a. mit Hilfe von Partner- und Gruppenarbeit sowie Rollenspielen. Alle Hörtexte sind auf Cassette aufgezeichnet: die entsprechenden Abschnitte sind mit dem Symbol gekennzeichnet.

Im Bereich Grammatik liegt der Schwerpunkt auf der Wiederholung wichtiger Strukturen, insbesondere solcher, die im technischen Englisch häufig sind.

Das Anspruchsniveau sowie das Vokabular der Schwerpunkte (*Manufacturing, Energy* usw.) ist so gehalten, dass dies für Lernende verschiedener technischer Fachrichtungen interessant, relevant und motivierend ist. Dass dies gelungen ist, liegt in hohem Maße auch an der Hilfe vieler Kollegen aus Schulen und (Fach-) Hochschulen sowie aus der betrieblichen Weiterbildung bei der Entwicklung und Erprobung des Materials. Der Verlag und der Autor sind ihnen dafür sehr dankbar.

Inhalt Foundation Course

Modules

UNIT 1

Manufacturing

A company visit

Wells Industries is a major supplier of semi-finished products to the automobile industry. The company wants to modernise its production plant and replace certain types of production machines. Production costs must also be reduced, by approximately 20%. A project group has decided to invite people from four machine tool companies to the plant in Stockton. Mrs Sue Wilkinson, an engineer in the project group, has written this letter to Hans Lindner, the German representative of ABA, an international manufacturer of machine tools and industrial robots.

Wells Industries

… and I was therefore pleased to hear that you will be able to come to Stockton on January 21st. As I mentioned on the phone, we are particularly interested in the X-2140 and in your IRB range of industrial robots. We will be expecting you at 10 a.m.

I understand you will be arriving by car from Birmingham. Enclosed are directions on how to find us – I am sure you will have no problems.

We are looking forward to seeing you.

Yours sincerely,

S. Wilkinson

S. Wilkinson

Focus Greetings and introductions

1 DIALOGUE

Hans Lindner meets the project group at Wells

Sue Wilkinson meets Mr Lindner at the main gate and takes him to her office where some of her colleagues are waiting.

Sue Wilkinson: Good morning Alan, morning Carol, good morning Ralph. Thanks for coming. I'd like you to meet Hans Lindner from ABA Germany. I think some of you have already spoken to him.

Carol: Hello, how are you, Mr Lindner? Yes, we've already spoken on the phone, I believe.

Mr Lindner: Yes, that's right. How are you?

Carol: Fine, thanks. And how are you?

Mr Lindner: Very well, thanks.

Carol: I hope you found us alright. I hear you came here by car.

Mr Lindner: Yes, no problem. Mrs Wilkinson sent me a map, so it was really no problem.

Ralph: Hello, Mr Lindner, very glad to see you. Did you have a safe trip?

Mr Lindner: Yes, thanks.

Sue Wilkinson: And this is Alan Waine. He's just been made our boss.

Mr Waine: Oh, come off it, Sue. Hello, Mr Lindner, very glad to see you. No, there is no boss. It's just me who reports to management.

Mr Lindner: How do you do? Pleased to meet you.

Sue Wilkinson: Now, how about some coffee?

PHRASES

Greetings and introductions

Hello. (It's good/nice to see you again.)
(formal:) Good morning/afternoon/evening.
(formal:) How do you do? – How do you do?
How are you? – Fine/Very well, thanks. And how are you?
This is .../Have you met ...?
(formal:) May I/Can I/Let me introduce you to ...
(formal:) I'd like to introduce you to .../I'd like you to meet ...
Nice/Pleased/Glad to meet you (, too).
Haven't we met before?
Did you have a good flight/trip/journey?
(informal:) How are things?
Do take a seat. Would you like/care for a cup of ...?

2 PRACTICE

Here are three jumbled-up conversations. There are three people in each. Work in groups of three and restore the conversations, then listen and check. After that, read or act them out.

Melba: Thank you. I've been looking forward to meeting you, too. How are things?

Ashmore: Hello, Mr Bauer. Sorry to keep you waiting. Did you have a good journey?

Ashmore: Oh yes. Thank you for sending it so promptly. Now, if you would like to follow me, please, I'd like you to meet Helen Krien first.

Schulz: Glad to meet you, Mr Larson.

Larson: How do you do? But haven't we met before?

Larson: Yes, that's right.

Fred: Oh, it's been a busy week, but things are easier now.

Schulz: Have we? ... Hmmm. Yes, I think you're right. That was probably last year in Lisbon. You were still working for Pirelli, weren't you?

Fred: Well, I think it's fine at the moment, but you'll have to see for yourself. Now, would you care for a cup of tea first?

Bauer: Yes, no problem. I hope you got my fax.

Melba: Thank you, that would be nice.

Fred: No, I haven't. But I've been looking forward to meeting you.

Larson: Yes, thanks. We left late, but the flight was O.K. And how are you?

Bauer: Good afternoon, my name is Bauer. I've come to see Mr Ashmore.

Schulz: Well, good to see you again.

Faulkner: Good afternoon. We've been expecting you. Do take a seat. Mr Ashmore will be with you in a minute. *(Pause)*

Melba: Yes, Paul was telling me about your new project. How's it going?

Paul: Hello, Fred, have you met Melba Macià?

Warner: Good morning, Mr Larson, nice to see you again. Did you have a good flight?

Warner: Fine, thank you. I'd like to introduce you to my colleague, Mr Schulz.

GRAMMAR

Simple past – present perfect – present perfect continuous

1	a	Mrs Wilkinson **contacted** Mr Lindner three days ago.
	b	Mrs Wilkinson **has** already **contacted** Mr Lindner.
2	a	Some time ago I **talked** to Anne about her new project.
	b	I believe we **have** already **talked** about the new project.

- The simple past (sentences 1a and 2a) is used to describe activities that were completed at a particular point of time or period in the past. The point or period of time is usually stated (*yesterday, a few minutes ago, in 1966* etc.), or is understood from the context.

- The present perfect (1b, 2b) is used to talk about activities that have happened, but we don't know or don't want to say when exactly, because it is not important. The fact that something has happened, not the time when it happened, is what matters.

3	**a**	Mr Lindner **has arrived**.
	b	The project group **has been waiting** since 10 o'clock.
4	**a**	I believe we **have** already **discussed** the new project.
	b	We **have been discussing** this project for long enough.

- The present perfect continuous (3b, 4b) is used to talk about activities which continued up to the present (and may still be going on).
- With the present perfect simple (3a, 4a) the result is important; with the present perfect continuous the emphasis is on the activity.
- Sometimes either form can be used, depending on whether the emphasis is more on the result or the activity.

---

3 PRACTICE

Complete the following sentences with the correct form of the simple past or present perfect.

1. We (modernise) ... our production facilities in 1996 and since then we (be able to) ... increase our efficiency by approximately 15%.
2. I (spend) ... three months in Spain when I was a student, but I (not be back) ... since then.
3. I (live) ... in Paris for the last three years. It (be) ... a great experience.
4. (send) ... Briggs ... the brochures yet? - I (not see) ... them yet but let me ask Margret. Here they are. Margret says they (arrive) ... yesterday evening but she (forget) ... to tell you.
5. I (try to) ... get through to John's office, but the line (be) ... busy all the time.
6. I'm glad to see you, Mr Calatravas. I (look forward to) ... meeting you.

---

4 TASK

Interview a partner, then briefly introduce your partner to the rest of the group. Find out:

- what kind of job your partner has or what she/he is studying, and how long she/he has been doing this
- what she/he did before
- where she/he lives and how long she/he has been living there
- whether she/he has any hobbies
- ...

Focus Describing a company

5 A profile of ABA

READING

Mr Lindner sent Wells some promotional material about ABA. Decide whether the verbs should be in the present, simple past or present perfect.

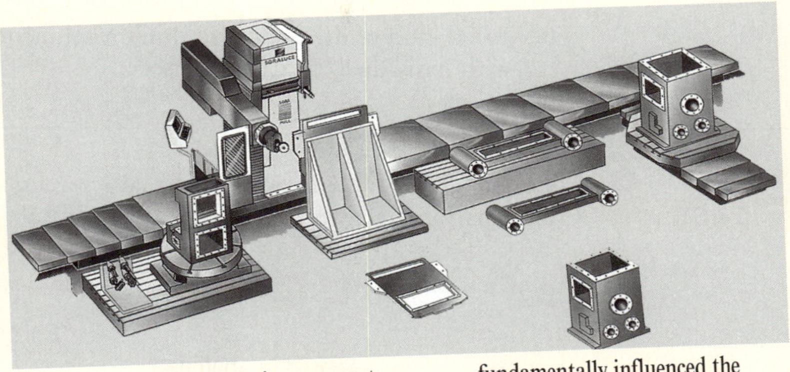

ABA first (begin) ... to develop its world-famous name for machine tools in the North American car industry during the late 1920s. We (be) ... among the very first pioneers in machine tool manufacturing to equip all our machines with separate electric motors, and by the 1940s ABA's machine tools were in use throughout the U.S.A. We soon (become) ... involved in overseas markets, and this (lead) ... to our expansion into Europe. Here, we (identify) ... new user needs and opened our plant in the Netherlands in 1952, and a plant in Scotland in 1957.

In the mid-1960s ABA concentrated on research and development, in response to demands for greater speed and reliability. We (improve) ... and expanded not only the product range, but also our service network – offering complete customer support wherever needed.

By the late 1970s, the climate of the machine tool industry was experiencing rapid change. Far Eastern competition fundamentally influenced the market, making purchase price a key issue. ABA (respond) ... by delivering a completely new product range, competitively priced with even further improved quality. The result was the revolutionary XL concept launched in 1984 which (influence) ... every ABA product since that time. It brings together every aspect of machine tool experience and design expertise, with the latest in development and production technology – in order to satisfy market demand for still higher quality, still lower costs.

We (always play) ... a leading role in the development of machine tools design and manufacture, right from the start. We (always meet) ... the needs of our customers and (respond) ... with technical innovation and new product introductions to meet ever greater demands – to the present day.

Today we (be) ... equipped to face the challenge of the 21st century through further investment in developing new products, through rapid answers to new customer priorities, and through greater efficiency in our plants and in our marketing services.

- Now answer the following questions.

1 Which innovation did ABA introduce in the 1920s?

2 Which developments marked ABA's activities in the 60s?

3 Why was the XL concept so important?

4 Which names and developments do you associate with the American car industry?

5 What kind of new product range do you think was introduced in the late 70s?

6 What kind of drives do you think machine tools had before they were equipped with individual electric motors?

---

6 TASK

The following table indicates the main developments of Wells Industries. Use the information given to write a text on the history of Wells Industries.

LOCATION

1965:	founded in Sheffield
1975:	new plant in Sheffield
1980:	moved to Stockton

MARKETS

First only domestic, now also western European countries (approximately 40%) and recently also eastern European countries (10% and increasing)

PRODUCT RANGE

1965 – 1979:	100% supplier to automobile industry
1979 – now:	70% supplier to automobile industry, 20% supplier to kitchen appliances industry, 10% other markets

WORKFORCE

1970:	480
1988:	700
1998:	620
2010:	expected to be 500

ECONOMIC SITUATION

70s and 80s:	very good
Last 5 years:	difficult
Prospects:	better in three years

NEW OBJECTIVES

Increased production safety

Better pollution control

ACTION TAKEN

Reduction of workforce continues; investment in new production technology continues; restructuring of decision-making procedures continues; steps towards lean production taken and to be increased.

Focus Describing production processes

7 Mr Lindner gives a presentation

LISTENING

After Mr Lindner has been shown the company, the project group meets again in the conference room for a presentation. Mr Lindner first explains different ways of organizing the production process in a company.

Make sure you understand the terms used in Figures 1 – 4, then listen for the main information: What sort of manufacturing systems do Figures 1 – 4 show?

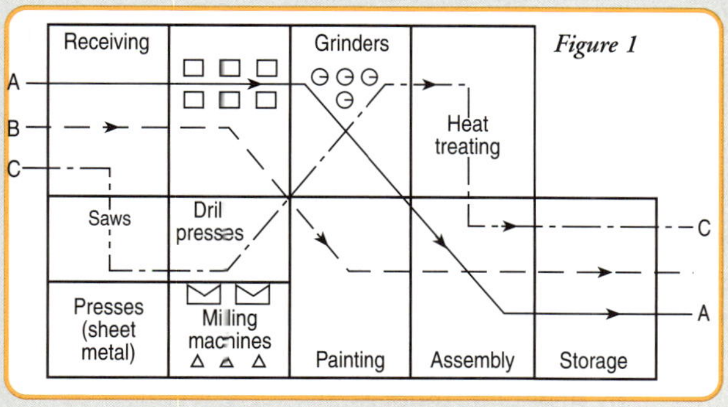

Figure 1

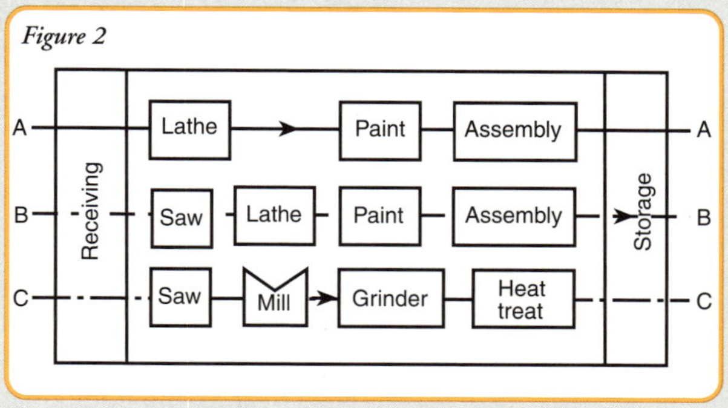

Figure 2

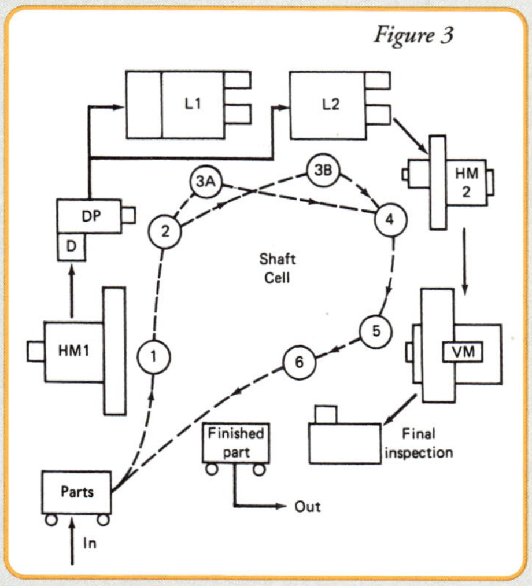

Figure 3

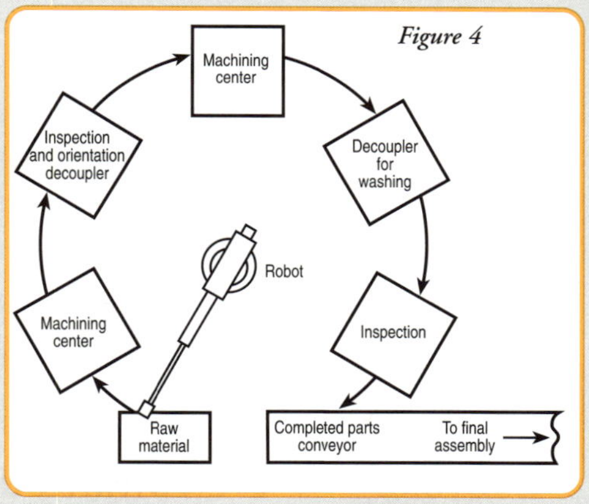

Figure 4

HM: Horizontal Milling Machine; DP: Drill Press
L: Lathe; VM: Vertical Milling Machine

- Now listen for detail. Take notes and answer the following questions.

1. How is the manufacturing process organized at Wells?
2. What is the main difference between a job-shop system and a flow-job system?
3. What are the advantages/disadvantages of a flow-job system?
4. What parts are manufactured in a flow-job system at Wells?
5. What is the main advantage of the cellular cell when compared with the flow-job system?
6. Which of Wells' departments would be ideal for the use of a robotic cell?

GRAMMAR

The passive

The majority of parts **are manufactured** in what is commonly called a job shop.

From the drill presses, part C **is brought** to the grinding machines before it **is heat treated, assembled** and **stored**.

- In the passive, a form of *be* is used with the past participle of the full verb (e.g. *heat, heated, heated/bring, brought, brought*). The form of *be* indicates the tense (e.g. *is, was, will be, has been*).
- The passive is used (especially in technical texts) when what is done is important, but not who does it. In the first sentence, for example, it is irrelevant (and possibly impossible to say) who manufactures the parts.

8

PRACTICE

Complete the sentences with one of the following verbs in the correct form.

can repair • improve • launch • equip • can change • need • replace • not give

1 Next year all the traditional lathes ... by new CNC models.
2 The new XL concept ... in 1984.
3 Different kinds of machines ... to complete the product range.
4 Over the last 15 years, the design of the machines ... so that they now meet the new safety regulations.
5 Now all our machines ... with sensors that stop the machines automatically.
6 Production ... over from one product to another almost immediately.
7 I don't know what happened, we ... any explanation for the changes.
8 Whether or not this machine ... depends on several factors.

CULTURE

Welcoming a visitor to a company[1]

Country	Situations	Physical greeting	Introducing self (I'm ...)	Introducing others (This is ...)
Germany	formal	handshake	normally only surname	normally title + surname
	informal	handshake	first name + surname	Mr/Mrs/Ms + surname
Britain	formal	handshake if offered	first name + surname	first name + surname
	informal	handshake at first meeting[2, 3]	first name (+ surname)	first name (+ surname)
USA	formal	handshake if offered[4]	first name	title + first name + surname
	informal	handshake[3]	first name	first name + surname
Japan	formal[5]	handshake	surname	Mr/Mrs/Ms + surname[6]
Scandinavia	formal	handshake	first name + surname	first name + surname
	informal	handshake[3]	first name + surname	first name + surname

[1] The table shows basic trends; behaviour may vary. [2] At first meeting, but probably not at later meetings. [3] Occasionally the greeting is a kiss (man – woman, woman – woman). [4] A man should always offer a handshake. [5] There is no informal way to introduce a visitor in Japan. [6] A title is used if the person is a medical doctor, professor or teacher.

UNIT 2

Energy

Renovating a house

PC-Extra is a small company specializing in importing and exporting computer accessories and software. The company is expanding fast, and the owner, Rajib Singh, has just bought a bigger house. A certain amount of renovation work is necessary, and Mr Singh's part-time secretary, Janet Buglass, is making preliminary inquiries about suitable companies.

One afternoon Mr Singh comes back from a meeting and finds a page torn out of the advertising section of the local newspaper, together with the following note from Janet.

Dear Rajib,

How did it go this morning?
I found this page in the D. Times. George says that Alpha are very reliable.
I've already phoned them, but they'll be busy until the end of June.
Is that too late?
We don't know about Home Improvement. Do you want me to contact them?
See you tomorrow.

Janet

Focus Saving energy

1 Make the most of your heating

READING

Mr Singh's house will need alterations to the heating system and other improvements to save energy, so the page from the local newspaper is of special interest to him. Read it and find the answers to the following questions.

1 How is wasting food and consumption of energy compared?

2 What is said about government regulations?

3 Which two means of saving energy are illustrated by the two pictures at the top of the pages?

4 What kind of companies are Home Improvement, Alpha and S. Fox? (See the advertisements.)

5 Would a similar text in a newspaper in this country mention the same facts?

GET READY FOR WINTER

... ALL AROUND THE HOUSE

NEIL'S PLUMBING & HEATING
232 Brimington Road, Tapton, Chesterfield S41 0ST
TEL: (01246) 203974
NO JOB TOO SMALL

SOLO GLAZING
White and Mahogany uPVC Windows, Doors and Conservatories
Broken window replacements
FREE QUOTATIONS, FREE ADVICE
For good quality at a fair price
Tel: Chesterfield 455444 (24 hrs)

S. FOX (Roofing) LTD
ALL TYPES OF ROOFING WORK UNDERTAKEN
Felt Roofing a Speciality
No Job Too Small
Free Estimates
Tel: Chesterfield 276870

MAKE THE MOST OF YOUR HEATING

There would be a public outcry if half the food bought by families in this country was thrown into the dustbin. Very rightly we would be reminded of all the people who are hungry, and particularly those in third-world countries. But many families throw away nearly half the energy they buy to heat the home, and do so with little thought.

Throwing away energy is perhaps even more important than throwing away food. Unlike food, most of our energy sources are irreplaceable. Where heating is by electricity, wasteful use can also cause damage to the environment by the excess emission of noxious gases from coal and oil-fired power stations. Therefore it not only reduces personal expenditure, but is also in the public interest to use energy for heating economically.

Advertising feature

Economic use of energy for heating does not mean that we have to be cold in our homes. It does mean we should use heat sensibly and obtain maximum benefit from the energy we buy, by ensuring that our homes are properly insulated.

Government is well aware of the need to use energy economically. Therefore during the past years it has gradually increased the standard of insulation which must be built into the walls and roofs of new homes. These are major areas of heat loss. As a result, walls and roofs of homes built today are nearly twice as effective in retaining heat compared with those built six or seven years ago and around three times more effective than those fifteen or more years old.

It is neither difficult nor expensive to improve the standard of insulation in most homes. Where there is heat loss due to badly fitting windows and outside doors, use draught stripping to stop the heat escaping and cold draughts coming into the rooms. Make sure you have four to six inches of insulation in the loft.

More heat escapes through the outside walls than from any other area, therefore these should also be properly insulated. This is usually a simple job in houses and bungalows built after about 1930. The walls of these homes have a two to three inch gap running up the middle which can be filled with a foam specially made for the purpose. The work is carried out from the outside and takes about half a day for a semi-detached house.

This is a job which should be done by a specialist, but the Cavity Foam Bureau at PO Box 79, Oldbury, Warley, West Midlands B69 4PW, telephone 0121-544-4949 can let you have a list of their members, all of whom are registered with, and approved by, the British Standards Institution.

So why not check that you are using energy for heating sensibly and that your home is properly insulated? The money you save is money in your pocket. You will also be conserving our natural resources and possibly helping to improve the environment.

2 PRACTICE

The words below are often used incorrectly. Check their meanings in your dictionary and select the right ones for the blanks. (The underlined words are used in the text above.)

to insulate - to isolate; to approve - to improve; to remind - to remember; sensible - sensitive; the environment - the surroundings; during - while; retain - contain

1 Our secretary ... me to call our new client. *(remind/remember)*
2 Wasteful use of energy will cause damage to our ... . *(surroundings/environment)*
3 Mr Swinson called ... you were out. *(during/while)*
4 You need to change the windows if you want to ... the heat in your house. *(retain/contain)*
5 We finally managed to ... the three most dangerous bacteria. *(insulate/isolate)*
6 Recycling will help ... the environment. *(improve/approve)*
7 ... my stay in Italy I did not find the time to visit our new supplier. *(during/while)*
8 Using our resources in a more ... way is more than important: it is a necessity. *(sensitive/sensible)*
9 Can you ..., what was their price? *(remember/remind)*
10 The head of my department ... of my suggestions to reduce the noise level in hall C. *(improved/approved)*
11 I agree that this is a nice house, but I don't like its ... . *(environment/surroundings)*
12 We need a vessel that will ... at least 1500 litres of braking fluid. *(retain/contain)*
13 The fire was started by an electrical wire that wasn't properly .... . *(insulated/isolated)*
14 This particular element is very ... to carbon monoxide. It detects very small amounts. *(sensible/sensitive)*

GRAMMAR

Adverbs and adjectives

1 This is a very **effective** heating system.
2 It is also very **reliable** and **economical**.
3 The new system has been **carefully** designed.
4 The insulation of outside walls is **especially** important.
5 The work can be done **very** quickly.

- Adjectives add information to a noun (sentence 1) or pronoun (sentence 2).
- Adverbs add information to a verb (sentence 3), an adjective (sentence 4) or another adverb (sentence 5).

3
PRACTICE

Decide whether an adjective or an adverb is needed in the following sentences.

1 The new house has been effective/effectively insulated against draughts.
2 There has been a gradual/gradually increase in standards of insulation.
3 As long as the house is not proper/properly insulated, a lot of heat will be wasted.
4 The house should not be unnecessary/unnecessarily warm.
5 The foam can be pumped into the cavities by means of special/specially equipment.
6 If you use heat sensible/sensibly, you can save a lot of money.
7 Good insulation is particular/particularly important in the loft.
8 The means of preventing heat loss must be analysed not only individual/individually but also in terms of how they affect one another.
9 I would strong/strongly recommend installing double-glazed windows.

4
TASK

Rajib Singh leaves Janet a message. He wants her to call Home Improvement and ask about their services, what they charge, and when someone could do a survey. He also wants her to work one full day that week.
Write the note, giving a reason why she should be in the office for one full day.

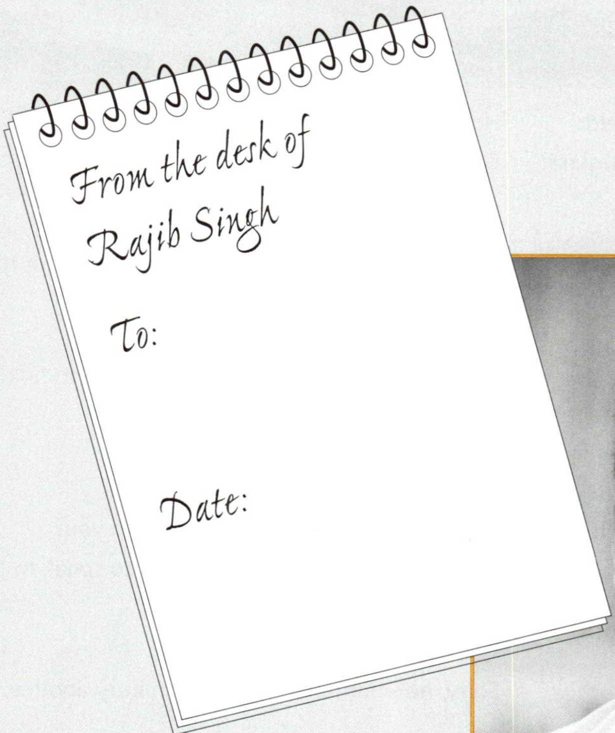

Focus Telephoning

5 Janet Buglass calls Home Improvement

DIALOGUE

Read and listen to the conversation. Why can't Janet speak to Mr Swinson?

Sally Field:	Home Improvement, how can I help you?
Janet Buglass:	Good morning, this is Janet Buglass of PC-Extra speaking. I've read your ad in *The Derbyshire Times* and would like to know more about the energy survey you're offering.
Sally Field:	Yes, you need to speak to Mr Swinson about that. Hold the line, please.
	…
Sally Field:	Sorry, the line's engaged. Do you want to hold or would you rather call back later?
Janet Buglass:	I'll call back later, thank you.
	30 minutes later
Sally Field:	Home Improvement, how can I help you?
Janet Buglass:	This is Janet Buglass again. I'd like to speak to Mr Swinson, please.
Sally Field:	Hold the line, please.
	…
Sally Field:	Sorry, he's not in his office. Let me try another number.
	…
Sally Field:	Mrs Buglass?
Janet Buglass:	Yes.
Sally Field:	I'm afraid he's just gone out. He should be back in about twenty minutes. Let me give you his extension number so you can call him in his office. The extension is 26.
Janet Buglass:	Thank you.
Sally Field:	You're welcome.

PHRASES

Telephoning

Hello, this is Rajib Singh speaking.	Benson Ltd, how can I help you?
I'd like to speak to ...	Who's speaking, please?
Can/Could I speak to ...?	Speaking.
Can/Could I call again later?	Sorry, the line is busy/engaged.
Can/Could you give me his/her extension number, please?	X is out of the office/not available right now, I'm afraid.
Can I leave a message?	(Would you like to) Hold the line, please.
Could he/she call me back?	I'll put you through to ...
	Shall I ask him/her to call you back?
	I'll get him/her to call you back.
	Can I take a message?

6

PRACTICE

Later Mr Singh calls Home Improvement to talk to Mr Swinson. What are the missing parts of the conversation below? Compare your versions with each other, then listen and check.

Sally Field: Home Improvement, how can I help you?
Mr Singh: ...[1]?
Sally Field: Sorry, but Mr Swinson is out for lunch. ...[2]?
Mr Singh: No thank you, I'd rather call again later. ...[3]?
Sally Field: He should be back around 2 o'clock.
Mr Singh: ...[4].
Sally Field: You're welcome, Mr Singh.
One hour later
Sally Field: Home Improvement, Sally Field speaking, how can I help you?
Mr Singh: I'd like to speak to Mr Swinson, please.
Sally Field: ...[5]?
Mr Singh: This is Mr Singh from PC Extra.
Sally Field: Oh yes, Mr Singh, ...[6] his office, just hold the line, please.
Mr Singh: Thank you.
She puts him through.
Mr Bicknell: Bicknell, hello.
Mr Singh: Uh, this is Mr Singh, could I speak to Mr Swinson, please?
Mr Bicknell: Just a second, I' ...[7].
Pause
Mr Swinson: Hello.
Mr Singh: Mr Swinson?
Mr Swinson: Speaking.

7

TASK

Work in pairs. In Pairwork Task 1 Rajib Singh (Partner A) phones Mr Swinson's office again. Partner B answers the phone.

A: Go to Pairwork Task 1, page 128. B: Go to Pairwork Task 1, page 130.

Focus Giving advice

8

LISTENING

Home Improvement has carried out an energy survey, and Mr Swinson phones with some suggestions. Mr Singh makes notes.

Listen for the main information. What two items are missing in Mr Singh's list?

- Now listen again for detail. What are the survey results for the boiler, pipes, radiators, windows, and walls/loft? What does Mr Swinson recommend, what will the initial costs be, and what effect will it have if Mr Singh follows the recommendation? Continue the table below on a piece of paper.

Replace boiler
Insulate pipes
Insulate walls and loft

Survey result	Recommendation	Initial costs	Effect
Oil-fired boiler	Replace with gas-fired one	Cost of oil-fired boiler plus £500	Annual saving of £60 – £80

PHRASES

Giving advice and making suggestions

Why don't you … ? I suggest (that) you … . I'd advise you to … .
How about buying a new boiler?
(I think) You should … .
I (would) recommend buying/that you buy a new boiler.
Have you thought of …ing … ?
If I were you, I would … .

9

TASK

Work in pairs. In Pairwork Task 2 Rajib Singh (Partner A) asks a friend (Partner B) for business advice.

A: Go to Pairwork Task 2, page 128. B: Go to Pairwork Task 2, page 131.

GRAMMAR

Conditional sentences

1	If you **replace** the windows by double-glazed ones, you **will save** a lot of money.
2	If half the food bought in this country **was thrown** away, there **would be** a public outcry.
3	If Mr Singh **had moved** into the house immediately, the renovation work **would have caused** problems.

- The condition can be fulfilled and the result is certain (sentence 1):
  if-clause: simple present – main clause: *will*-future
- The condition cannot be fulfilled or is improbable (sentence 2):
  if-clause: simple past – main clause: *would/could/might* + infinitive
- The condition was not fulfilled in the past:
  if-clause: past perfect – main clause: *would/could/might* + *have* + past participle

10
PRACTICE

Complete the following sentences.

1 If we don't stop the emission of CFCs, ... *(the ozone layer/be destroyed)*.
2 If we had spent as much on developing alternative sources of energy as we spent on developing nuclear energy, ... *(alternative energies/already become cost-effective)*.
3 I would ride my bike to work if ... *(I/not live so far away)*.
4 *(I/install solar panels)* ... if the weather was better in this area.
5 Would you install solar heating if ... *(the government/support it financially)*?
6 If *(my friend/not help me)* ..., I would not have been able to carry out the renovation so quickly.

CULTURE

Telephoning[1]

	What time to call on weekdays		What people say when they answer the phone.	
	at work	at home	company receptionist	at home
Germany	8.00 – 12.00, 14.00 – 16.00	8.00 – 22.00	company name	surname
Britain	10.00 – 12.00, 14.00 – 16.00	8.00 – 22.00	company name + "How can I help you?"	telephone number or "Hello"
USA	8.00 – 18.00	7.30 – 21.00[2]	first name + surname + company name[3]	"Hello"
Japan	10.00 – 11.30, 13.00 – 17.00	8.00 – 22.00	company name	surname, "moshi moshi"
Scandinavia	8.00 – 11.30, 12.00 – 16.00	8.00 – 22.00	company name	surname
Spain	9.00 – 13.00, 15.00 – 17.00	9.00 – 22.30	company name	name or "Si?" or "Diga" or "Digame"
France	8.00 – 12.00, 14.00 – 18.00	9.00 – 22.00	company name	"Allo?"
Italy	9.00 – 12.00, 14.00 – 16.00	8.00 – 22.00	company name	"Pronto" or "Hello"

[1] The table shows basic trends; behaviour may differ.
[2] In the evening always ask if you are interrupting dinner.
[3] Many companies now have voice-mail systems instead of receptionist.

UNIT 3

Electricity

Lift design

Jan Olafson, an engineer for the Swedish lift company Hansen, met Mr Bretton, a sales representative for the British company Howland Electric, at the Hanover fair. Some weeks after the fair, Mr Olafson receives the following letter.

Dear Mr Olafson,

I was very pleased to meet you at the Hanover fair and was impressed by your company's innovations in lift design.

I have to apologize for still not being able to send you the material on lift drives you expressed an interest in. Due to a major restructuring of our company, the R&D department has not yet found the time to attend to your enquiry. But I can assure you that we are doing everything possible to supply you with the information as soon as possible.

Enclosed you will find some brochures, which I hope will be of interest to you.

Please also give my apologies to your colleague, Mrs Larsson – she has been very patient with me. I am enclosing the article I promised to send her.

Yours sincerely,

John Bretton, Eng

Encl.: Brochures

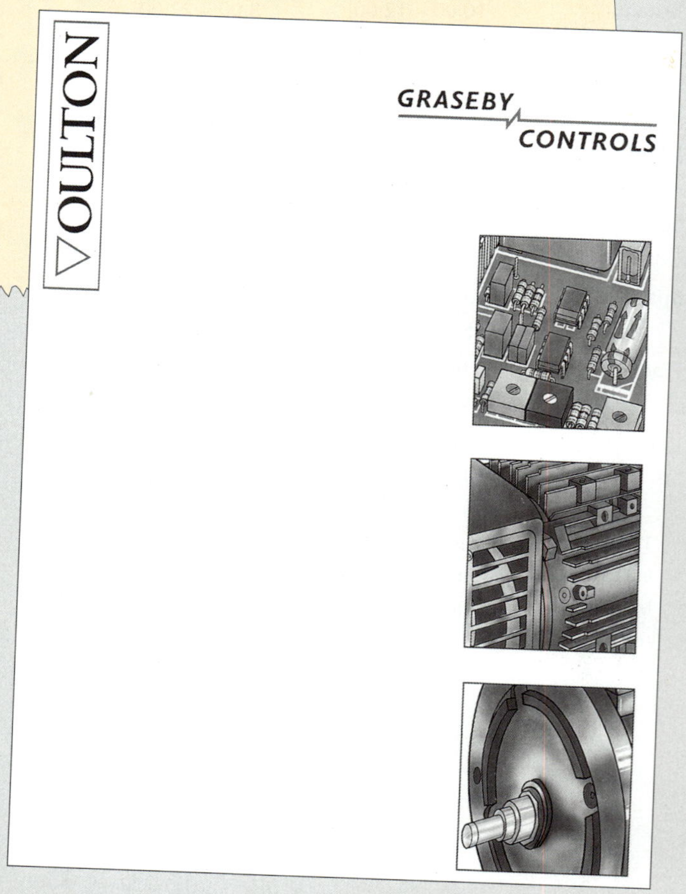

Focus Arranging a meeting

1 Mr Olafson phones to fix a date for a visit

LISTENING

Mr Olafson will be in England during the first week of June and would like to visit Mr Bretton at Howland. The diary below shows some of Mr Olafson's commitments. Listen to the phone conversation: when do Mr Olafson and Mr Bretton agree to meet?

	MONDAG	TISDAG	ONSDAG	TORSDAG	FREDAG	LÖRDAG
9 – 11		9.00 Pinter (Sheffield) Tel: 94368 Ext 16 ↓	9.45 Claris??	9.30 Cobel (Mr Rees) (Brochures) ↓	9.45 Price & Portman	
11 – 13	11.30 Airport Manchester	13.00		11.30		
13 – 15	14.00 Peter Taylor ↓		16.00 Perkins, Liverpool		15.00 Windsor Hotel Mrs Pyne ↓	15.15 KLM 348 16.45
15 – 18	17.00?				17.15 Theatre?	

PHRASES

Making an appointment

- I was wondering whether we could meet (on ...).
- Would you be free/available on ... ?
- How/What about Tuesday?
- Would you rather meet on Friday or on Thursday?
- I'm afraid I will be busy/out of the office then.
- I'm awfully sorry, that won't be possible, but what about ... ?
- That sounds good./That's fine.
- Would 2 o'clock suit you?
- So it's 2 o'clock on Friday 16th then. I'm looking forward to seeing you.

2
TASK

Work in pairs. In Task 3 Mr Ian Toll (Partner A), an engineer with the British company Flex Plc, which produces optical fibres, receives a phone call. The caller is Mr Grundland (Partner B), an engineer with the Swedish company Transcom, a company in the telecommunications industry. The two men have already met at a trade fair.

A: Go to Task 3, page 128. B: Go to Task 3, page 131.

PHRASES

Apologizing and reacting to apologies

I am (awfully) sorry that I'm late/sorry for being late.
(formal:) Please accept my apologies for the delay.
(formal:) I do apologize/I owe you an apology for …
Don't worry about it.
Never mind, that's quite all right.
That's quite all right.

PHRASES

Giving reasons

(formal:) Due to *(formal:)* Owing to *(formal:)* On account of	a major restructuring of the company, I am unable to …
Because As Since	the company is being restructured, I am unable to …

3
TASK

Write a fax to rearrange a meeting.

A few days after Mr Olafson's phone call, an accident happens in one of the companies he wants to visit, and he has to rearrange the schedule for his visit to England. He faxes Mr Bretton explaining the reason for the change and apologizing for the inconvenience. He proposes another week for the meeting. Write the fax, using expressions from the boxes above.

GRAMMAR

Past perfect simple and continuous

Mr Olafson and Mr Bretton **had already** met at the Hanover fair before Mr Olafson came to England.
Mr Olafson **had been trying** to phone Mr Bretton for several hours before he decided to send him a fax instead.

- The past perfect is used to express the fact that an event in the past *(past perfect)* was completed before another event in the past *(simple past).*
- The conjunctions *after* and *when* are often used with the past perfect.
- The past perfect continuous indicates that the first activity had been going on for some time.
- Continuous forms are only possible with verbs that describe an activity, not with those that describe a state.

4

PRACTICE

Complete the following sentences with the correct past tenses.

1 Mr Olafson … *(leave)* work early last Monday. He … *(feel ill)* the whole day and finally … *(decide)* it would be better to see a doctor.

2 When Mr Bretton … *(leave)* his local pub last Wednesday night, he … *(cannot find)* his car. He then … *(realize)* that he … *(park)* it in a no-parking zone. The police … *(take)* it away.

3 When Mr Bretton … *(come)* to work, he … *(find)* the fax which Mr Olafson … *(send)* him.

4 After he … *(read)* the fax, Mr Bretton … *(telephone)* Mr Olafson. They … *(arrange)* to have a meeting on Wednesday.

5 Before Mr Olafson … *(join)* Hansen, he … *(already work)* in the electronics industry for several years. He … *(look)* for a more interesting job, and that's what Hansen … *(offer)* him.

6 Mr Olafson … *(fly)* from Stockholm to Manchester and then … *(take)* a taxi to his hotel. He … *(make)* the reservation two weeks earlier.

7 Mr Olafson … *(forget)* to bring the brochures that he needed. So he … *(ring)* his secretary and … *(ask)* her to send them.

8 Mr Bretton … *(already have)* two meetings that day when he … *(meet)* Mr Olafson at half past one.

9 Mr Bretton … *(try)* to get in touch with a colleague all day. Then he … *(remember)* that she … *(be)* on holiday.

10 After he … *(take)* his dog for a walk, Mr Bretton … *(go)* home and … *(watch)* a film with his wife.

11 Howland Electric … *(sell)* mostly in the domestic market before they … *(attend)* the Hanover fair. They … *(be)* therefore quite interested in the new Swedish contacts.

Focus Describing a technical product

5 A Graseby brochure

READING

Hansen is bidding for a contract for the lift work in a new high-rise office building in Sweden. The company's normal supplier of motors may not be able to supply appropriate ones; this is why Mr Olafson is in Britain. One of the companies that might be able to help is Graseby Controls. He has been sent a Graseby brochure by Mr Bretton with some preliminary information.

Read for the main information: what did Mr Olafson find out about the points on the right?

1. the company's experience in manufacturing electric motors?

2. the main advantages of their motors?

3. concrete examples for the use of their motors?

GRASEBY CONTROLS

OULTON

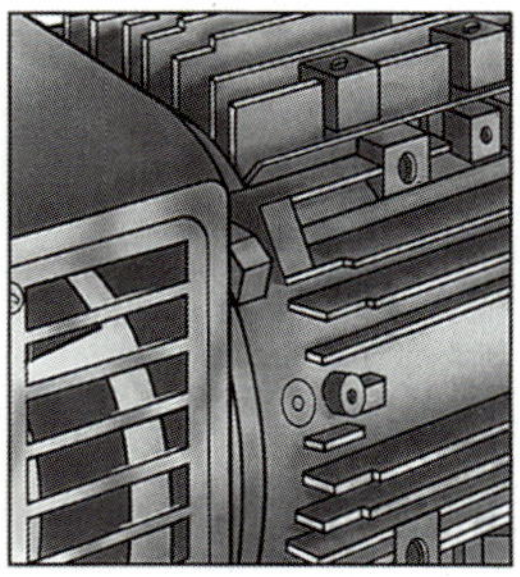

Graseby Controls, a member of the Graseby Group plc, is one of the market leaders in design and manufacture of innovative and durable drives and controls system.

Graseby Controls' unique experience and expertise provides a first-class level of service for all industrial applications, and has earned the company a reputation as providers of total, cost-effective solutions to all motion control problems.

FEATURES & BENEFITS

OULTON

The Oulton Switched Reluctance Drive is the result of twenty years research and development by the leading electrical and magnetics experts at British Universities, combined with the extensive industrial drives and manufacturing expertise of Graseby Controls.

This combination of original design and engineering know-how has produced a world first in an outstanding range of high performance switched reluctance drives for industrial applications.

Discover the benefits of high performance ... with Oulton.

TOP-CLASS PERFORMANCE

Zero regulation (speed change) from no load to full load. Steady-state stability excellent. Response time only 40mS. Speed range 100:1. Repeatability better than + 1 rpm.

HIGH EFFICIENCY

Up to 91% total system efficiency, maintained down to low speeds, for example 88% at half speed.

SIMPLE MOTOR

Totally Enclosed Fan Ventilated, Hoseproofed as standard. No routine maintenance.

ADVANCED CONTROL ELECTRONICS

Constructed for easy access and installation. Uses fast-acting IGBT transistors. Includes soft-start and forward/reverse switching.

CASE STUDY

The following case study illustrates just one example of how the Oulton High Performance Drive System can actually pay for itself through massively increased efficiency.

OULTON DRIVE GIVES FIVE MONTHS PAY BACK

Last year a district hospital appointed an independent engineer to estimate possible savings by increasing the efficiency of their drive systems used on fans in their coal-fired steam boilers.

The four boilers, with a combined output of over 100 million lb's of steam every year, use draught fans to aid combustion. The fans were driven by fixed speed 22kW, 1500 rpm induction motors. In order to make a comparison, a 20-week trial on one of the boilers was carried out, using an Oulton Drive System; comprising the drive and a 15kW, 1500 rpm switched reluctance motor type PLD 160M/1 with the standard IP55 specification enclosure. Over the 20-week trial the Oulton System showed substantial savings. A further test by the Department of Energy showed that the Oulton would make a projected saving of some 22,400kW of electricity over a 12-month period – in all a saving of over £10,000.

The capital cost of converting the boiler to the Oulton System was £4,000, providing a simple pay-back period of just 20 weeks.

In addition to the financial saving, the Oulton System provided future advantages and benefits. These included extending periods between routine cleaning due to increased efficiency of combustion; the soft-starting characteristics of the Oulton, coupled with the ability to have extended periods of reduced speed operation caused far less wear and tear on the drive transmission and fan.

GRASEBY CONTROLS

- Now read for specific information. Look at the following notes that Mr Olafson took as he read, and decide whether they are correct or incorrect. If they are incorrect, correct them.

Oulton » a subsidiary of Graseby Controls

- *The company cooperates with other manufacturers of electric motors.*
- *Oulton drives are mainly used in the manufacturing industry.*
- *Case study based on university research – quite impressive.*
- *In the test trial the original motor was replaced by a 22KW Oulton motor.*
- *The test took 12 months to complete.*
- *Tests showed little wear on draught fans.*

Focus Problems and solutions

6 LISTENING

Mr Olafson visits Howland Electric

Mr Olafson and Mr Bretton meet to discuss Hansen's needs. Listen for the main information and explain what Mr Olafson's problem is.

- Now listen for specific information and answer the following questions.

1 Why is Hansen keen on winning the contract?

2 What were the original contract requirements?

3 What are the contract requirements now?

4 How will the lift cars be different now?

5 How does Mr Olafson explain the new demands on the lift motors?

7 PRACTICE

Complete the extracts from the dialogue in 6, using some, but not all of the following phrases (two words per space). Then listen and check.

acceleration time • a challenge • a contract • contract requirements • emergency regulations • energy consumption • had solved • lift contract • motor supplier • new demands • response time • very disappointed

1 **Mr Bretton:** Yes, I remember you mentioning it. Quite a big contract, I imagine.
Mr Olafson: Yes, it is. And quite ...[1] for us.

2 **Mr Olafson:** We were doing pretty well up to that point. You know that one of the ...[2] was to come up with a design which would have at least 15% lower running costs.
Mr Bretton: Yes, and you mentioned problems with your ...[3].

3 **Mr Olafson:** But for the running costs we'll have to start again. First of all, the individual cars will have to be bigger and thus heavier. ...[4], however, will have to be higher if we don't want to increase the response time to a call too much. This, of course, will increase ...[5] for each individual car. What it means for the whole system is a different question. But the real snag is that it places completely ...[6] on the lift motors.

4 **Mr Bretton:** But that was already a problem with the original design, wasn't it?
Mr Olafson: Yes, but one we ...[7]. We'd got round the problem of fire regulations as well.

PHRASES

Describing problems and solutions

The problem/snag/challenge is …
We've met/hit a problem/snag/difficulty.
The problem has been solved.
Our solution is/was to …
We've managed to get round the problem.

8 TASK

Work in pairs. In Pairwork Task 4 you will find information about a new idea that solved a technical problem. Use the information to prepare an explanation of the problem and how it was solved. Present your result to your partner, using the expressions above.

Make sure you know the meaning of the words *nut, bolt, fasten* and *sleeve* before you start.

A: Go to Pairwork Task 4, page 129.
B: Go to Pairwork Task 4, page 132.

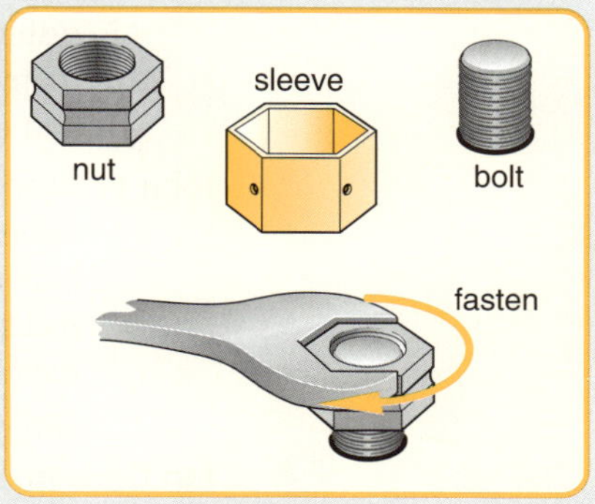

9 PRACTICE

Match the opposites. The words on the left are in the text on pages 28 – 29.

1	increased	a	fragile
2	extensive	b	conventional
3	durable	c	limited
4	industrial	d	standard
5	substantial	e	negligible
6	unique	f	domestic
7	innovative	g	decreased

Focus Describing a graph

10 DIALOGUE

Mr Olafson describes typical lift-travel activities

To explain to Mr Bretton the requirements for the lift motors Mr Olafson first describes typical lift-travel activities throughout the day.

First look at the graph and decide what you think the answers to the following questions are. Then check with the text.

1 Which line is probably for up traffic, which for down traffic?

2 What might the reasons be for the fluctuations and peaks?

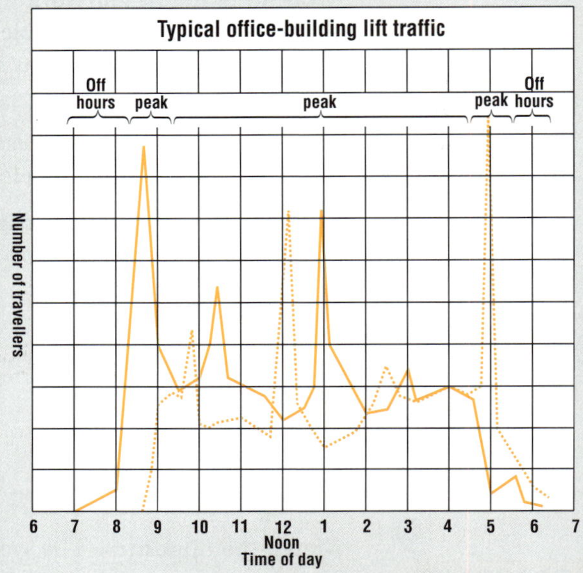

Jan Olafson: I suppose you realize that large lift systems require fairly sophisticated control systems if you want them to run effectively and economically.

John Bretton: Yes. But to be honest, I'm not too familiar with the details. As I said, we've dealt mainly with industrial applications.

Jan Olafson: I'll explain it very briefly then. Now, the thing to understand is that if there are several lifts serving one building, the controls must decide immediately which car to send to a call. And it may very well be that there are several calls simultaneously from different floors. Let's say there are calls from the 3rd, the 14th and the 22nd floor. The one from the 22nd wants to go down, the one from the 14th wants to go up. Somehow, the control system must deal with it and decide which car – let's say there are 4 cars – goes to which floor.

John Bretton: Yes, I understand.

Jan Olafson: And there is something else. As you can imagine, travel demands vary substantially throughout the day.

John Bretton: Yes.

Jan Olafson: Now, let's have a look at the chart. It shows typical travel activities for an office building, with the continuous line, here, this one, indicating the up travel and the broken line showing the down travel. Now, up travel starts around 8 o'clock – actually this chart is for a building in America, in Sweden we tend to start earlier.

John Bretton: What time would that be?

Jan Olafson: Probably at 6.30 to seven. Anyway, you see that the curve rises steadily until it hits a maximum shortly before nine, when the curve begins to fall rather sharply. Between nine and half past twelve the curve fluctuates a little ...

John Bretton: ... with a minor peak around 10.30. Why's that?

Jan Olafson: That's a typical coffee-break situation.

Any further questions?

English for Technical Purposes – Textcassette
Mit diesen Texten – von geschulten englischen Sprechern gelesen – üben Sie Ihr Hörverständnis und Ihre Aussprache. Die Cassette enthält Aufnahmen aller Hörverstehenstexte und eine Auswahl der abgedruckten Texte. Hören Sie doch mal rein!

Studium kompakt:
Studieren in Großbritannien/Nordirland
Eine wertvolle Hilfe für Studierende: Diese kontrastiv angelegte Einführung bereitet detailliert und umfassend auf den Studienaufenthalt auf den britischen Inseln vor:
- Bildungssystem
- Landeskunde
- Sprache und Spezialwortschatz
- Ausführliche Übungen und Tests
- Praktische Hinweise und Sprachhilfen

Der Studienführer ist in deutscher Sprache verfasst, einzelne Abschnitte sind auf Englisch.

Weitere Angebote auf der Rückseite.

Bestellkarte

Hiermit bestelle ich zur Lieferung wie umseitig ausgewiesen (per Nachnahme oder Bankeinzugsverfahren). Preis zuzüglich Versandkosten.

☐ **English for Technical Purposes – Textcassette**
Bestellnr. 25055 ◇ 35,80 DM

☐ **Studium kompakt: Studieren in Großbritannien/Nordirland**
160 Seiten, Bestellnr. 31187 24,80 DM

☐ **Englische Grammatik für Erwachsene: 99 Steps to Grammar and Usage**
240 Seiten, Bestellnr. 35956 27,90 DM

☐ **Technical Grammar und Vocabulary**
144 Seiten, Bestellnr. 20436 29,80 DM

☐ **Wortschatz Technisches Englisch**
160 Seiten, Bestellnr. 27457 26,90 DM

☐ **Handwörterbuch Technik:**
Deutsch-Englisch/Englisch-Deutsch
2 Bände. 1000 Seiten, Bestellnr. 494454 108,— DM

☐ **Deutsch-Englisch**
Bestellnr. 494446 59,80 DM

☐ **Englisch-Deutsch**
Bestellnr. 494438 59,80 DM

Gewünschtes bitte in dem entsprechenden Kästchen deutlich ankreuzen. Alle Titel können auch durch den Buchhandel zum Ladenpreis ohne Versandkosten bezogen werden.
Preisstand 1.1.2000. Änderungen vorbehalten. ◇ = Unverbindliche Preisempfehlung.

Englische Grammatik für Erwachsene – 99 Steps to Grammar and Usage
Ein Nachschlage- und Übungsbuch, mit dem Sie Ihre Grammatikkenntnisse festigen und ausbauen können.
Jedes Kapitel beginnt mit einem kleinen Test. Sie überprüfen selbst, wie viel Sie schon wissen, und erfahren, was Sie noch üben sollten. Die folgenden Regeln sind klar und einfach formuliert. Beispielsätze verdeutlichen die Merksätze. Auf die Unterschiede zum Deutschen wird ausdrücklich hingewiesen.
Mit Hilfe der sich anschließenden Übungen lernen Sie, die Regeln anzuwenden und auch die Gesetzmäßigkeiten und Besonderheiten der fremden Sprache zu erkennen.

Technical Grammar and Vocabulary
A Practice Book for Foreign Students.
Der Band liefert den Wortschatz, die Redemittel und die typischen Strukturen des allgemeinen technischen Englisch, ohne sich in spezifischen Fachgebieten zu vertiefen. Die vielfältigen Vokabel- und Grammatikübungen sind nach Themen geordnet, z. B. *energy, transport, robots, factory work, civil engineering, electricity, automotive technology.* Ein Lösungsschlüssel liegt bei.

Wortschatz Technisches Englisch
Thematisch geordnet vermittelt dieses Lernwörterbuch technisches Vokabular. 75 alphabetisch geordnete Rubriken behandeln die gebräuchlichsten Begriffe (vor allem aus dem Maschinenbau), bei denen es zu Unklarheiten, Verwechslungen und Fehlern bei der Übersetzung kommen kann. Der Anhang enthält Übungen (mit Lösungen) sowie ein deutsches und ein englisches Stichwortverzeichnis zum gezielten Nachschlagen einzelner Vokabeln.

Handwörterbuch Technik
Deutsch-Englisch/Englisch-Deutsch.
Das glossarartig konzipierte Fachwörterbuch wendet sich an Studierende ebenso wie an Techniker und Ingenieure in der Praxis. Die Ausgabe umfasst je Band 45.000 Einträge, von denen eine Hälfte zur allgemeinen Technik gehört, die andere verteilt sich auf 15 verschiedene technische Fachgebiete.

Cornelsen online
Besuchen Sie uns im Internet: Neben dem Online-Katalog mit allen Titeln aus unserem Verlagsprogramm und der Möglichkeit zum Bestellen finden Sie auch aktuelle Informationen. Also surfen Sie mal herein:
www.cornelsen.de

Bitte in Druckschrift komplett ausfüllen

Name / Vorname

Straße / Hausnummer

Postleitzahl / Ort

Datum / Unterschrift

Ich bestelle zur Lieferung:

☐ per Nachnahme (Preis zzgl. Versandkosten und 6,50 DM Nachnahmegebühr)

☐ per Bankeinzug (Preis zzgl. Versandkosten)

Kontonummer

Bankleitzahl

Name des Kontoinhabers

Datum / Unterschrift des Kontoinhabers

Bitte freimachen

Antwortkarte

Cornelsen

Cornelsen Verlag

14328 Berlin

John Bretton: Oh, yes. Of course.

Jan Olafson: And the curve really shoots up for lunch-time. Now, after lunch-time it goes up and down a bit with a minor increase in up-travel demand around three o'clock, and you can see the sudden drop towards half past four when people start leaving the building.

John Bretton: Yes.

Jan Olafson: Now, this travel pattern is very important for the design of the control system. It must be able to anticipate the travel demand and send cars to waiting positions accordingly. And also, and this is where you come in, there are periods of very high demands on the lift motors.

John Bretton: Yes, I see what you're getting at ...

- Mr Olafson doesn't want to sound like an instructor who knows everything. What phrases does he use to soften what he's saying and create a more equal, cooperative relationship?

---

11 PRACTICE

Match the words/phrases and graphs. For some graphs several expressions may be used.

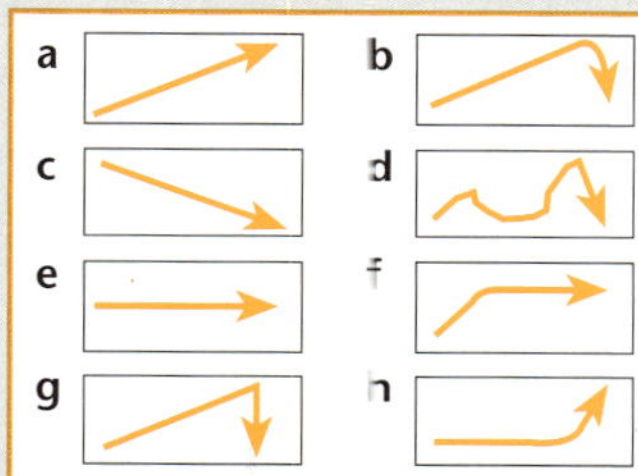

1 to remain constant
2 to shoot up / to shoot down
3 to increase / the increase
4 to decrease / the decrease
5 to flatten out
6 to go up / to go down
7 to level off
8 to drop / the drop
9 to rise / the rise
10 to hit a maximum/minimum
11 to fluctuate / the fluctuation
12 to fall / the fall
13 to climb / the climb
14 to recover / the recovery
15 to peak / the peak
16 to plunge / the plunge

The following adverbs are often used as modifiers: *sharply, gradually, suddenly, steadily, drastically.*

---

12 TASK

Work in pairs. In Pairwork Task 5 you will find a graph (Partner A: manufacturing output in 1998/9; Partner B: a task for your partner). Prepare a description of your graph, then describe it to your partner. He/She must draw it according to your description.

To describe the graph you may need some of the following phrases in addition to those above.

The graph/chart shows ...
The horizontal/vertical axis represents ...
The horizontal/vertical axis is divided into ... units/sections (, each representing ...).
The horizontal/vertical axis has a scale from ... to ...

A: Go to Pairwork Task 5, page 129.
B: Go to Pairwork Task 5, page 132.

CULTURE

Working hours and holidays[1]

	Working hours per week	Working day	Lunchbreak
Germany	37.5 +	7.30/8.30 – 15.30/17.00	30 – 45 min, 12.00 – 13.30
Britain	36 – 38	8.00/9.00 – 17.00[2]	60 min, 12.00 – 2.00
USA	40	9.00 – 17.00	12.00 – 13.00
Japan	41	9.00 – 17.00	12.00 – 13.00
Scandinavia	37.5	8.00 – 16.00	11.30 – 12.00
Spain	40 +	8.00 – 15.00/16.00	13.00 – 14.00/15.30
France	42	8.00 – 20.00	12.00 – 14.00
Italy	38	8.00 – 18.00	60 min, 12.00 – 14.00

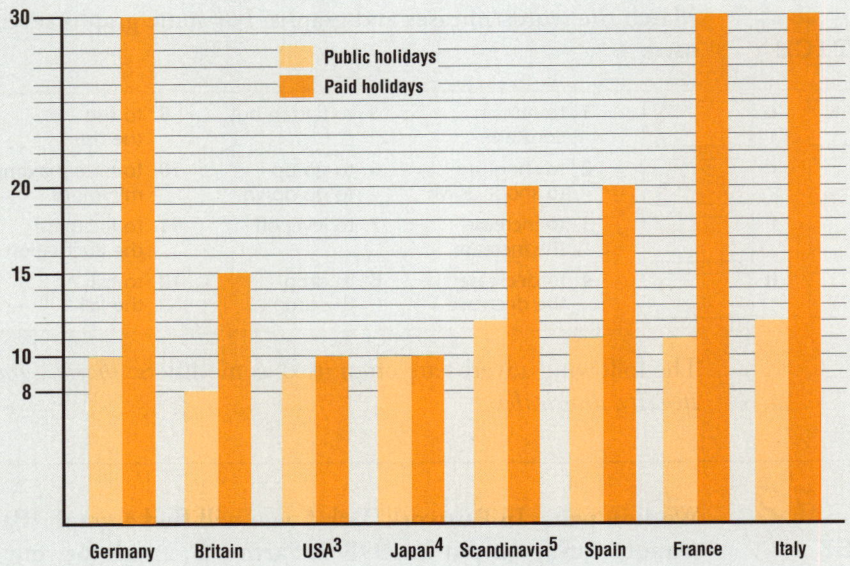

[1] The tables show basic trends only; there may be small variations.

[2] Factories start at 8.00, offices at 9.00.

[3] Higher level employees get 20 or more days after at least 10 years with company.

[4] 10 days at beginning, then 1 day per year up to maximum of 20 days.

[5] 25 days for people over 60.

UNIT 4

Architecture

Architecture and art

Richard Watts and Tom Fielding are architects working for the English company David Shield & Partners. The company is preparing a bid for a new railway terminal to be built in Birmingham and wants to send the two men on a research tour to the Continent to see some of the latest designs in airport and railway terminals. One day when Tom Fielding is out of town, Richard Watts finds an interesting article on a building in Switzerland in a magazine, and sends him this fax.

DAVID SHIELD & PARTNERS

Fax

Attention of: Tom Fielding
From: Richard Watts
No. pages incl. top sheet: 5
Time/Date: 11.15am 11th April

Please telephone us immediately if you do not receive the number of pages indicated above.

Dear Tom,

I'm sending you a copy of an article about a house in Switzerland. It's not really our line, but I'd like to see it. How about a stopover in Zurich? What do you think? Give me a ring as soon as possible.

Cheers,

Richard

P.S.: Great job you did on the Surrey job. Time to celebrate.

TASK

The fax is clearly written from one friend to another. What changes would you expect if Richard had written to a colleague who was not a close friend? Rewrite the fax accordingly.

Focus Describing a building

1 READING

The Zumikon Residence in Zurich

Below is part of the article Richard sent Tom. Read the article and answer this question: In what sense(s) is the Zumikon Residence two homes/buildings?

On a gently sloping hill above Lake Zurich stands a house which Gwathmey Siegel designed for a large family and an equally large modern art collection that the family regards as an integral part of its domestic surroundings. The commission required the architect to fit an ambitious program onto a site governed by the strict Swiss building code, which not only limits height and mass, but also requires the construction of a stick mock-up, so that townspeople can judge the design for themselves. From the street, the south-facing front facade, standing two and a half storeys high, echoes the retaining walls characteristic of Zurich. Rooms are grouped into two main wings, connected by a dining room. A raised cylinder accommodates the reading room part of the library and, one floor up, the master bedroom. This space, the only round element in the design, acts as a hinge between the relatively public and more private areas of the house.

The living room is the most striking space in the house. Twenty-four feet high with a glazed south face, the room is dominated by a powerful painting by Richard Long and an impressive open fire-place.

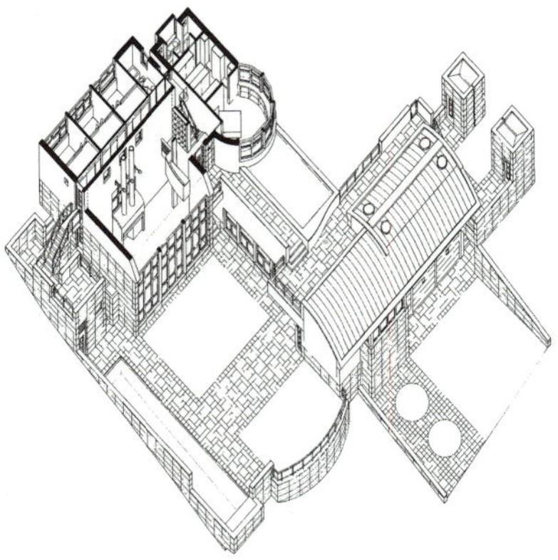

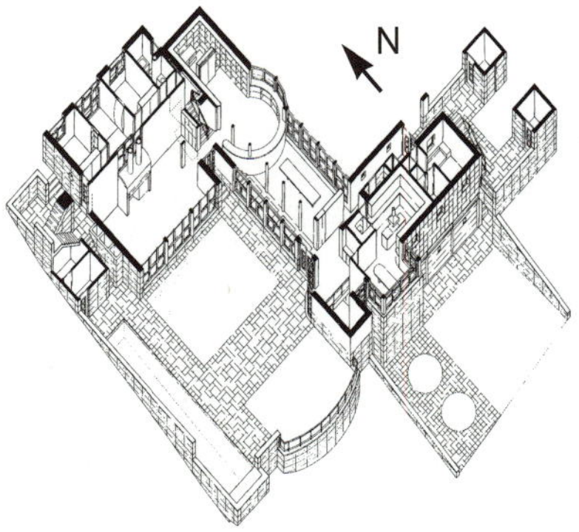

- Now work in pairs and identify the dining room, the library and reading room, the master bedroom, the living room.

  When you have finished, explain your solution to the rest of the group.

2 PRACTICE

Check the dictionary for the meaning of the following expressions and explain what is meant.

1 ... an ambitious program *(line 6)*
2 ... requires the construction of a stick mock-up *(line 9)*
3 ... echoes the retaining walls *(line 13)*
4 ... accommodates the reading room *(line 15)*
5 ... acts as a hinge *(line 18)*
6 ... the most striking space in the house *(line 21)*
7 ... the room is dominated by a powerful painting *(line 23)*

Focus Arranging a visit

3 Richard Watts writes a letter to Zurich

READING

Tom Fielding agreed that it would be interesting to go to Zurich, so Richard Watts writes the following letter to the Swiss office of the European Architects' Association in Zurich.

Read and note down what they would like to do in Zurich and why the Zumikon Residence is of interest to them.

DAVID SHIELD & PARTNERS

ARCHITECTS AND CONSULTANTS

Ashtree House
12 Marley Road
Huddersfield HD4 5DD
England
Telephone: 01484 4121280
Telex: 52316 DBSP A
Fax: 01484 4221887

10 April 20-

European Architects' Association
Grohestrasse 71
CH-8051 Zurich
Switzerland

Dear Sir or Madam,

The American Journal "Architectural Record" recently carried a short article on the Zumikon Residence that Gwathmey Siegel planned and built some years ago. The article caught my immediate interest as the architects seem to have been able to integrate a rather large building well into its natural surroundings.

David Shield and Partners are primarily concerned with works in the public sector. Our involvement has so far mainly been in airport and railway terminal constructions as well as in council office buildings. However, some of the constraints we have to meet are, it would seem, similar to the ones that were so successfully met in the Zumikon Residence.

A colleague and I are planning a research tour to some European cities and we would very much like to take the opportunity to visit the Residence. Do you think such a visit would be possible? We would, of course, be very pleased if an exchange of ideas with someone from Gwathmey Siegel could also be arranged.

Apart from our travel dates – we would be on the Continent in the week of 12th – 19th June –, we don't yet have any fixed commitments; so we are still very flexible. Could I perhaps ask you to give me a call or send me a fax saying whether you think it would be possible to arrange a visit?

I look forward to hearing from you.

Yours faithfully,

Richard Watts
Richard Watts, BSc, Ceng, MIStructE

PHRASES

Letter writing

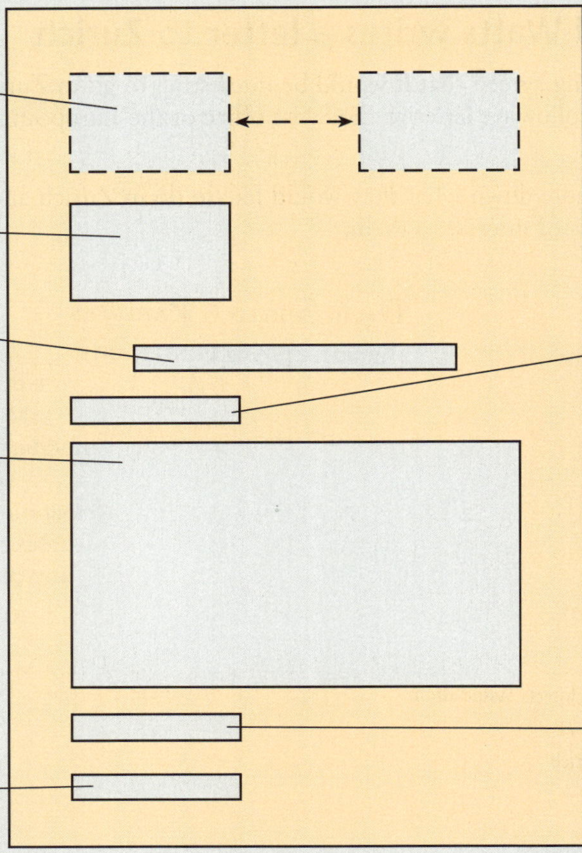

4 TASK

Write a letter from Mr Schlueli of the European Architects' Association to Mr Watts. Include the following points.

– Thank Mr Watts for his letter and interest in the building.

– Say that the owner and the architect agree and will join Mr Watts for the visit.

– Say that Mr Watts should phone for details.

5 LISTENING

Richard Watts phones Mr Schlueli to fix a date for their visit to the Zumikon building. Listen for the main point: what date and time do they agree on?

● Now listen again for further details. Answer the following questions.

1 What does Mr Schlueli say about the owner of the Zumikon building?

2 Which hotel does Mr Schlueli recommend, and what does he say about it?

3 Who is going to make the hotel reservation?

6 PRACTICE

Complete the following sentences from the dialogue. Use one word per space. Then listen again and check.

1 **Mr Schlueli:** I …[1] you're calling about your visit to the Zumikon building.

Mr Watts: Yes, that's right. We're very …[2] to you for having made a visit possible. I hope we won't be taking too much of your time.

Mr Schlueli: No, … … …[3], it's a pleasure. I'm looking forward to showing you the building with the architect. He was very much involved in its design and enjoyed it very much.

2 **Mr. Schlueli:** Actually you'll be talking to the owner as well. He's a very interesting person and very proud of the building. So he'll be very …[4] to show you around.

Mr Watts: I'm very …[5] to hear that. Now, what day would be …[6] convenient? We're very …[7].

Mr Schlueli: The owner said Tuesday … …[8] him best. And that's fine with the architect, and me, too.

3 **Mr Watts:** Thank you very much again.

Mr Schlueli: My …[9]. Goodbye, Mr Watts.

PHRASES

Expressing and responding to thanks

Thank you very much for … .
I really appreciate your help in this matter.
We're very grateful for your help in this matter/in making this possible.
We're very grateful to you for …ing … .
I'd like to express my thanks to you for … .
Oh, you're welcome.
I am glad I could help.
It's a pleasure./My pleasure.
Don't mention it.

FOCUS Taking someone on a guided tour

7 A tour of the house

LISTENING

Unfortunately the owner of the Zumikon building and the architect had to cancel the meeting at short notice. So, after viewing the house from the outside, Mr Fielding and Mr Watts return to Mr Schlueli's office, where they are shown the plans of the building and given a guided tour on paper.

Listen first for general information. Say whether the following statements are true or false, and correct any that are false.

1. The house has a total of approximately 1,000 feet of floor space.
2. The "tour" starts on the first level, then they go upstairs.
3. Different sorts of works of art are usually on display at the house.
4. One reason for the size of the house is that the family is large and they often have guests.

- Now look at the plans and the layout of the house; the "tour" starts at point 15. Listen again and identify the parts of the building.

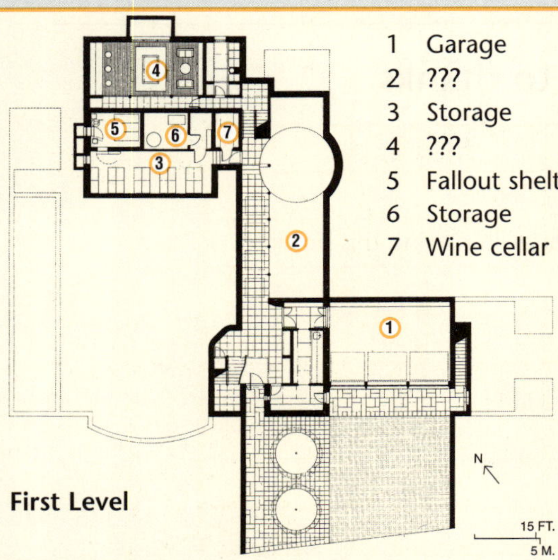

1 Garage
2 ???
3 Storage
4 ???
5 Fallout shelter
6 Storage
7 Wine cellar

First Level

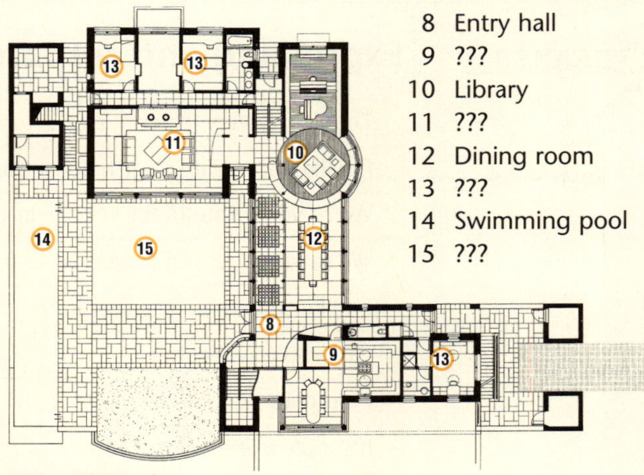

8 Entry hall
9 ???
10 Library
11 ???
12 Dining room
13 ???
14 Swimming pool
15 ???

Second level

PHRASES

Showing someone around a building/factory

Let's first look at the general layout of the plant.
Our offices occupy a total floor space of 1,200 square feet.
If you would like to follow me this way.
Assembly is carried out in building A; building B houses the power plant.
This area is where orders are processed.
I'm afraid you'll have to put on/wear your hard hats now.
I'm sorry to have to tell you that you are not allowed to take pictures.

Focus Architecture and engineering

8 An interview with an architect about his work

READING

At the European Architects' Association office, Richard Watts read an interview with Santiago Calatrava in a copy of the journal *Architectural Record*. Santiago Calatrava is a very influential architect of public works who, at the time, was designing a new railway station for the city of Lisbon. In the interview he also talks about the railway station he built near Lyon and about more general aspects of his profession.

Read part of the article and do the tasks that follow.

AR = Architectural Record, SC = Santiago Calatrava

AR: In many countries they tend to distinguish between architecture and engineering. What is your opinion on that?

SC: I think it is really out of fashion to make such distinctions. I believe in an understanding of architecture similar to that of Frank Lloyd Wright. Wright had a fair amount of technical training before he went to work with Louis Sullivan. His precision in the use of materials, his respect for technology: to me those are the keys to his poetics. Both engineering and architecture serve the art of construction. My ideal is to make works of pure engineering that are created by the soul of the artist.

AR: Do you see any changes taking place in the way architects and engineers are educated? Do you think the training they receive is still too different?

SC: The training of engineers is still strongly directed toward the analysis of phenomena through mathematical models. They don't learn to interpret nature in a purely visual way, by drawing. Many schools are now trying to bridge the gap by teaching engineers both the history of engineering and the history of art. I can't stress enough how important it is for an engineer to realize that the profession is not just part of a technical process, but also part of the culture of the 19th and 20th centuries. An engineer can make a cultural contribution. The Golden Gate Bridge is as much an expression of our culture as the architecture of, say, the skyscrapers in New York.

AR: Would you agree that the late 20th century's "emancipation" of engineering is not just a matter of attitude, but also of technology – in particular the arrival of computers.

SC: Yes, very much so. Computers liberate the engineer from the technical analyses which have always been difficult and time-consuming tasks. They give the engineer more time to concentrate on the nature of the work itself. They also allow more freedom to think about how to construct an object and to find innovative solutions. Engineering has always been a search for ingenious answers, and technology can help. Engineering is the art of the possible.

AR: Whether it is stations, airports, bridges or schools, 99 percent of your work comes from winning competitions, and almost all those competitions are conducted by public bodies.

SC: Hmm. Yes, most of my work is from public commissions. It's time for authorities to become sensitive to the value of public works. Their responsibility is not to make public works as cheap as possible, but as beautiful as possible. Art is not just museums, but also streets: What we use and touch, what we walk on and drive in every day. Bridges, stations, airports: they are built to serve society. Good architecture is good for everybody. It shows that a building can be not only functional, but also a symbol of what a community wants to be, of how it wants the outside world to see it.

AR: And the TGV station and airport connection in Lyon are an example of that.

SC: The city asked for a building that would serve as a regional symbol. The unusual thing was that they wanted not a political or institutional building, but a large public one.

AR: The bridge you designed in Barcelona is another example, isn't it? A public works commission that when it was completed became a symbol of the city.

SC: Bridges are not just standard objects determined by cold calculation. Some are ugly, some are beautiful. A bridge is more than a way to cross an obstacle. As soon as we recognize that, we are on the way to discovering that it is a work of art. The Golden Gate Bridge is one of the masterpieces of the 20th century, and I mean not just the art of construction, but all the visual art.

First look at the structure of the text: in which order does Santiago Calatrava talk about the following topics? Indicate the line numbers where the relevant sections start and finish.

1 Computers have changed the way engineers work and made them freer.
2 The training of engineers is still very technical.
3 Public works can and should be works of art as well as something functional.
4 Architecture and engineering belong together as two parts of the creative process of construction.
5 Authorities have a duty to create beautiful buildings.
6 Engineers should learn that their work is part of our wider culture.

- Now read the text again, take notes, then write a summary.

GRAMMAR

The definite article

1 a **Engineering** and **art** are related. They are both expressions of man's **creativity**.

b **Engineering** and **architecture** serve **the art** of construction.

2 a **Architects** and **engineers** depend on one another.

b **The architects** who designed this building worked closely together with the owners.

- The definite article is not used before abstract nouns and indefinite plural nouns (1a, 2a) except when the noun is qualified by a word or phrase so that it refers to a particular type or group (2a, 2b).
- A number of places do not have an article after a preposition if the institution or purpose of the place is meant. An article is required if a specific, physical building or place is meant:
  - to/at/from school/university/college/work
  - to/in/from town
  - to/into/in/from church
  - in/to/into/out of prison/hospital/bed
  - at/from home

9

PRACTICE

Decide if an article is needed or not in the following sentences. (If both are possible, explain why.)

1. It is time for … authorities to become sensitive to … value of public works.
2. Frank Lloyd Wright had great respect for … technology.
3. … history shows that … architecture has always played an important part in every society.
4. … computers give the architect more time to concentrate on … nature of … work itself.
5. … students should be taught … history of engineering as well as … history of art.
6. The well-known architect Norman Foster went to … university in England and the United States of America.
7. Bridges, stations and airports are built to serve … society.
8. When you return from … work, could you stop at … church in Radnell Street and check when it was built.
9. The Golden Gate Bridge is as much an expression of … art as … architecture of the skyscrapers in New York City.
10. We've just installed … computers we bought last week. They will give me more time for … creative work.

PHRASES

Expressing an opinion, agreeing and disagreeing

In my opinion …	I can only agree with you there.
The way I see it …	I think Robert is right.
To my mind …	I agree with you in principle, but …
It seems to me that …	I find it difficult to agree with you here because …
From my point of view there is …	I can see your point of view, but …
It is essential that …	Well, I think to a certain point Gareth is right, but …
I'd like to emphasize that …	I'm afraid I (totally) disagree with you …

10 TASK

Discussion. Divide into groups of four or five and discuss the questions below. Decide on a spokesperson who will later report the results of your discussion to the class.

1 What differences do you see between architecture and engineering?

2 Would you agree that it is important to see a building as an object of art rather than just space that serves a purpose?

3 Should the public be involved in the decision-making process of the design of a building or should that be left to the experts?

CULTURE

Etiquette of meetings[1]

	Formal meetings	Informal social meetings		
	Arrive on time?	**Usual place?**	**Be on time?**	**Present if at private home?**
Germany	yes	restaurant	+ 10 min	flowers
Britain	yes	pub	+ 20 min	flowers/chocolates (lady)
USA	yes (+ 5 min)	bar/pub	+ 15 – 20 min	flowers/wine/candy
Japan	yes	pub	+ 20 min	sweets, fruit, alcohol
Scandinavia	yes	—[3]	yes	—[3]
Spain	yes (+ 15 min)	restaurant	+ 30 min	flowers[4]
France	yes[2]	restaurant/bar	yes[2]	flowers/wine/chocolate[5, 6]
Italy	yes	restaurant/ at home	+ 10 – 15 min	flowers (lady)/wine (man)[6]

1 The table shows basic trends; behaviour may vary.

2 In the south of France you can be 15 mins late for a formal meeting, 30 for an informal one.

3 Informal meetings are rare.

4 Invitations are rare.

5 A present is not just appreciated, but expected.

6 Something typical from the visitor's country would also be appropriate.

UNIT 5

Recycling

Attending a congress

Redstock Auto Engineering is a US company that supplies parts to a US car manufacturer. John Warner, one of Redstock's employees, has seen the advertisement below in a magazine.

Read the advertisement quickly and answer the following questions.

1 Who is the advertisement aimed at?

2 Is it a trade fair that is being advertised or something more than that?

3 When and where is the event?

4 Does it say what costs are involved?

EUROPE'S LEADING AUTOTECHNOLOGY EVENT

AUTOTECH brings together the most influential and respected automotive engineers from around the world. Running over four days, around 40 seminars will be presented at the Congress by some 130 authors, all experts in the technologies under discussion.

Essential developments in materials, engine technology, design, power transmission, whole vehicle technology, environmental aspects, legislation – and virtually every other relevant topic of value – will be open to discussion.

Organized by the Institution of Mechanical Engineers, the Congress structure continues to lead world opinions in technological advances in the automotive industry.

National and multi-national companies exhibit state-of-the art automotive technology from all over the world.

Organized by Centre Exhibitions, a Division of National Exhibition Centre Limited, this Exhibition, coupled with the Congress Programme, ensures AUTOTECH remains the single most important meeting place for automotive engineers.

AUTO Tech 20..

12 – 15 November
National Exhibition Centre
Birmingham England

INTERNATIONAL BUS TRUCK & CAR PRODUCT AND MANUFACTURING TECHNOLOGY EXHIBITION & CONGRESS

Focus Getting someone to do something

1 LISTENING

Who is going to Birmingham?

John Warner talks to two colleagues about the Autotech, Brian Tusa and Lisa Bronowski.

Listen for the main points; answer the following questions.

1 What does John want?

2 Is he successful? Why (not)?

- Now listen again for further information; answer the following questions.

1 What does Brian say about the project he is working on and how does John react to it?

2 Describe Brian's responses, from his first reaction to his refusal to go. How does he react at each stage of the conversation?

3 Brian suggests that John should ask Lisa. How does John react?

4 How would you describe the way John introduces the topic to Lisa?

5 How does Lisa respond to John's suggestion?

6 Do you think that Lisa will go? Give reasons.

7 How would you describe John's personality? What is his relationship to Brian, Lisa and Robert?

PHRASES

Asking someone to do something, refusing politely

I'd like you to …
Could you please …?
I was wondering whether you could …
How would you feel about …?
Don't you think you could …?
I'd really appreciate it if you could …
I'd love to, but I really don't see how …
I'm awfully sorry, but …
Unfortunately I won't be able to …
It sounds very interesting, but …

2 TASK

Work in pairs. In Pairwork Task 6, Partner A wants someone to do something for him/her; Partner B refuses to do it.

A: Go to Pairwork Task 6, page 129. B: Go to Pairwork Task 6, page 132.

GRAMMAR

The future with "going to", "will" and the present continuous

A:	I'm **going to** leave work early today, it's been a hard day.
B:	**I'll come** with you. Perhaps we can get the 4.43.
A:	Yes. By the way, I think **I'll** probably **come** in early tomorrow.
B:	Right. My wife is **flying** to Boston tomorrow, so I'**ll come** by car.

- *Going to* is preferred when we talk about future activities that we have already decided to do.
- *Will* is preferred when a decision about a future activity is made at the moment of speaking.
- *Will* is also used to make a prediction.
- The present continuous can be used to talk about firm future plans and arrangements.
- You can often use *going to* or the present continuous for plans and arrangements: with *going* to the focus is on the intention, with the present continuous it is on the fact that the arrangements have been made.

3 PRACTICE

In the following dialogue put the verbs in the list into the kind of future tense you think fits best. (The verbs are not listed in the correct order.) When you have finished, read the text with your partner, then listen to the cassette and compare it with your version.

call • talk • go (2) • stay (2) • attend • rent • think • add on • send • be • not be

Robert: Hi Lisa, this is Robert.

Lisa: Hi Robert. How's life?

Robert: Fine, thanks. I hear you …[1] a conference in Britain this fall. Is that true?

Lisa: Yes, looks very much like it.

Robert: Great timing!

Lisa: What do you mean? I'll have plenty of time to get my projects finished by then.

Robert: No. I mean November. Great time for England, isn't it?

Lisa: Funny, very funny.

Robert: Just a joke. Actually, I just love England, even in the fall or winter. In fact, I …[2] there myself in September.

Lisa: Are you, where …[3]?

Robert: Actually not far from Birmingham, a little place, about, say 25 kilometres north of Birmingham. We …[4] a house for a week, my wife and I, while I have some work to do in the Birmingham area.

Lisa: Great idea!

Robert: Yes. What about yourself ? Have you decided where you …[5] yet?

Lisa: Yes, I …[6] at the Excelsior. Quite expensive but apparently very close to the Exhibition Centre.

Robert: Yeah, it's right next to it, but not my taste. I prefer places in the country and take a bus or train or hire a car to get to the conference. That way I get to see something of the country.

Lisa: Hm, sounds good. I …[7] about it. Maybe I …[8] a few days.

Robert: Yeah, do that. There are some great museums but also some beautiful countryside. I …[9] you some material if you like. But listen, why I'm calling you is to invite you to a party on Saturday.

Lisa: Oh, that's nice.

Robert: Yeah, I'm 40 this weekend.

Lisa: Oh, congratulations! Let me see. I was going to see a movie and have dinner with Jacky, but I guess I can change that. I …[10] you tomorrow to confirm. OK?

Robert: Not tomorrow morning as I probably …[11] here then. The afternoon …[12] better.

Lisa: Good, I …[13] to you later then, bye.

Focus Attending a presentation

4 Lisa Bronowski attends a presentation at the Autotech

LISTENING

At the congress Lisa attends a presentation on the prospects of recycling cars. Before you listen, exchange your information on the following questions.

1 What was traditionally done with cars when they reached the end of their lifetime?

2 What parts of a car present the least problem for recycling, and what percentage of the car's weight do they represent?

3 Which auto-parts present the main recycling problem? Has that changed over the last decades?

- Listen to the presentation and answer the following questions:

1 How does the recycling of cars compare with the recycling of other products?

2 What is "light waste", and what is done with it?

3 How many cars were dismantled and recycled over a period of 5 years?

4 What did BMW learn with its pilot plant?

5 What kind of information was recorded in the environmental data bank?

5 BMW's recycling project

READING

Read the second part of the presentation. What success did the company achieve? What were its future aims?

The cars for the pilot plant in Landshut came mainly from BMW dealerships but BMW also processed a number of Mercedes under contract. The dismantling process began by draining all fluids to avoid contamination of materials at a later stage. The dry car was then dismantled by removing the doors etc. to have easy access to the rest of the vehicle. Next the interior was removed and segregated into its various subassemblies. This was followed by parts that could be removed relatively easily such as bumpers, windows, side panels etc.

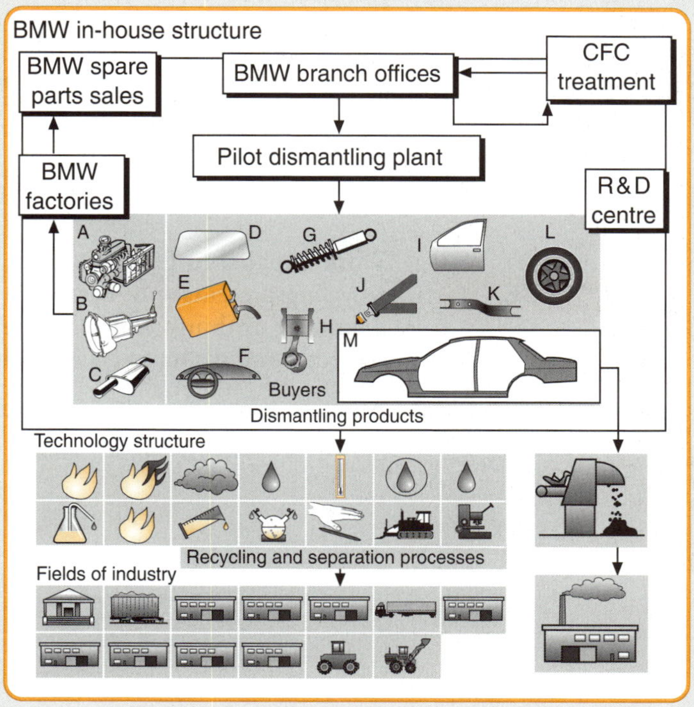

Then the main mechanical components such as engine, gearbox, axle, starter, generator and water pump were taken out, and finally the wiring was stripped out to leave a bare "clean" body shell. The main mechanical components were sent to BMW factories to be reconditioned and rebuilt for resale while the bare body was shredded.

Stripped of its non-ferrous materials, the pulp was then sent off to the steel mill while the non-ferrous metals were smelted and used again. The plastic, glass, rubber, cables and tires were separated and also shredded before they were returned to the raw materials supplier or other industries for reuse. A certain amount of residue always remained, which was either sent to landfill or – if the calorific value was high enough – used as fuel for the generation of electricity.

The recycling initiative reduced potential landfill by a considerable amount. A typical model left only 140 kg of waste compared with approximately 260 kg before the pilot project. This is a reduction of close to 50 percent, but engineers were sure that the figure could be reduced by another 50 percent within a ten-year period.

They also felt the process of dismantling could become more and more cost-effective. For example, by reducing the number of different kinds of plastics, the time necessary to dismantle a dashboard, identify and segregate the different components and introduce them into the appropriate recycling system would be reduced considerably. At the time a dashboard consisted of various different plastic materials, for example of foam, of skin etc. These combinations of materials were very difficult to recycle. The initiative showed that engineers needed to think in terms of recycling as well as in terms of engineering.

PHRASES

Clarifying a misunderstanding

You said that 500,000 vehicles were recycled annually in Germany.
Actually, what I did say was that …
What I meant to say was …
I'm afraid there's a misunderstanding. What I said was …
I'm afraid I haven't made myself clear enough. …
I don't think that's quite what I said. …
I'm sorry, I made a mistake there. The figure is …

---

6 TASK

Work in pairs. In Pairwork Task 7 Partner A is a member of the audience at the recycling presentation, Partner B is the presenter. A has taken notes, and there are a number of things which he/she is not quite sure whether he/she has understood correctly and wants to check with the presenter.

A: Go to Pairwork Task 7, page 130. B: Go to Pairwork Task 7, page 132.

PHRASES

Speculating and expressing possibility

I imagine letter A refers to …	It is possible that …
I suppose part B shows the …	I think …
I suspect …	Letter A could be …
I would say that …	Part F might be the …

7
TASK

Role-play a conversation after the recycling presentation.

The day after the presentation Lisa talks to Fred Michnum and Ralph Olafson, other visitors at the conference, who were not at the recycling presentation. Lisa tells them about the presentation. The presenter distributed a leaflet showing the in-house structure of the BMW recycling project (see page 49). As the diagram is not very clear they are trying to interpret which car parts (A – M) are shown in the diagram. Role-play the conversation in groups of three, use the phrases above.

8
PRACTICE

Decide which prefix to use with the following verbs in order to complete the text below. (The verbs are in the right order.)

The prefix *dis-* is often used to give a word its opposite meaning, for example: *to disappear*. In contrast, the prefix *re-* usually indicates something done "again", for example: *to reconsider*. The prefix *mis-* often indicates something done in the wrong way, for example: *to misunderstand*.

to claim • to appear • to place • to cover • to connect • to condition • to adjust • to integrate • to charge • to fill • to organize • to manage

In a world with only limited supplies of raw material, efforts must be made to …[1] valuable resources from consumer articles once they are no longer used. The old type of scrapyards in which whole cars are simply shredded are finally beginning to …[2]. They are being …[3] by modern recycling plants which …[4] some of the valuable materials such as steel, copper and aluminium.

The first thing to do before the car is dismantled is to …[5] the battery to avoid accidental fires. Certain parts, such as the gearbox, the starter and the engine are sent to special factories which check whether they can be …[6].The plastic industry is …[7] its marketing strategies and is beginning to talk about plastic bags that …[8] after a certain time, and the toy industry is making more use of batteries that can be …[9]. In some countries certain drinks have always been sold in glass bottles which can be …[10]. Now even washing powder comes in refill-containers. Farming techniques are also changing, farmers are …[11] their methods and producing "green" produce. If the new environmentally friendly farms are not …[12], they have a very good chance of becoming commercially successful.

- What other prefixes do you know? Make a list and explain their meaning.

Focus Giving a presentation

9 TASK

In groups of three to four read the two "Basic rules to observe when giving a presentation". Do you agree with them? Give reasons. Formulate at least three more rules. When you are finished, compare your results.

> Basic rules to observe when giving a presentation
>
> 1 Try not to start your presentation with an apology for something you think is not quite right with your presentation.
>
> 2 Keep eye contact with your audience, not with just one person. (In class, do not keep eye contact only/mainly with the teacher.)

PHRASES

Giving a presentation

Introduction:

I'm going to talk about …

My topic today is …

Sequencing:

Let me start with …

First of all …

Firstly, …, secondly, …, finally …

One final word about …

Let me now turn to …

Drawing attention:

As you can see …

I'd like to draw your attention to …

What's particularly interesting here …

The diagram shows …

Ending:

Last but not least …

To sum up …

As I have said before …

As I hope I have shown …

10 TASK

Give a presentation.

The diagram shows how the recycling system for cars used to operate in Japan and the kind of changes that will be made in the future. Prepare a presentation of the past and future systems. Observe the rules formulated in section 9 and use the expressions in the box above.

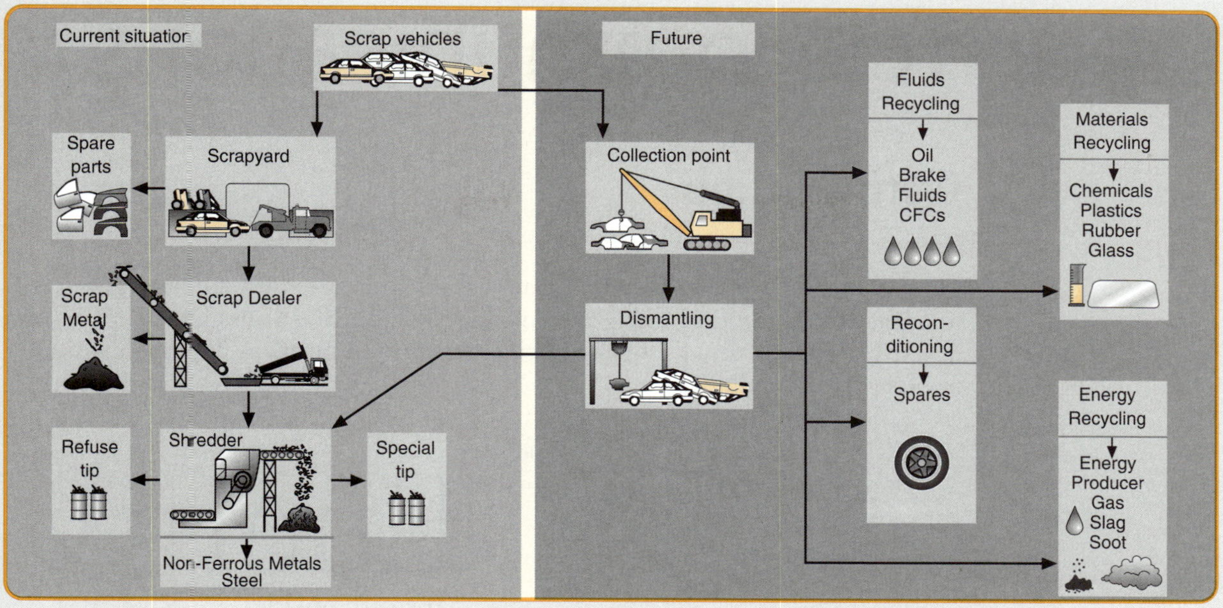

CULTURE

Dress and topics of conversation[1]

	Dress		Topics of conversation with foreign visitors	
	Normal working days casual = 1 formal = 5	When receiving foreign visitors at work casual = 1 formal = 5	Acceptable topics (examples)	Unacceptable topics (examples)
Germany	2 – 4	4	world politics, sport, travel	touchy parts of country's history, salaries
England	4 – 5[2]	5	each other's companies, local culture, travel	politics, religion, health, money
USA	2 – 4	4 – 5	history of the countries, sport, travel	money, religion, politics
Japan	4 – 5	5	each other's countries (food, culture, sights etc)	touchy parts of country's history, criticism of anything relating to the visitor
Scandinavia	1	1 – 5[3]	each other's companies, sport, free time, culture	
Spain	4	4	sport, food, history, general politics	salaries, religion, personal politics
France	2 – 4	2 – 4[4]	travel, local culture & history, places to visit, own country	salaries, religion and politics
Italy	2 – 4	2 – 4[4]	each other's companies, local culture, world economy	politics, religion, salaries

1 The table shows basic trends; behaviour may vary.

2 Men: jacket and tie; women: jacket (no jeans!)

3 More formal if someone is trying to dress like the visitor.

4 More formal dress is chosen if the visitor is being met outside the company buildings.

UNIT 6

Telecommunications

Getting a job

Carol Mistadou, a young Frenchwoman, has just completed her studies in telecommunications engineering and is now looking for a job. She would like to go abroad if possible, so has started studying foreign job advertisements in English-language newspapers and online.

Analysts **Up to £35,000** **Ipswich**

Eastern Power & Energy Trading (EPET) manages Eastern Group's energy portfolio, co-ordinating and optimising the Group's generation, risk management, gas and electricity sales businesses. Commercially, EPET are continuing their rapid growth with the establishment of a number of European offices.

The combination of well managed integrated activities and substantial new business has attracted some of the most commercially astute team players around. If you have the ability to thrive under pressure in a dynamic environment you could join them in any one of the following areas; Business Planning, Risk Management, Gas Trading, Commercial Development or the newly formed Transnational Trading. All areas offer superb opportunities for an exciting career with excellent prospects.

Your skills may be suited to a particular area, but in general you will be a numerate graduate, preferably in business, maths, statistics, engineering or an economics related subject, with at least two years' experience within a highly commercial role. Analytical ability is obviously a pre-requisite for developing an understanding of EPET's markets, 'modelling' and forecasting the implications of varied projects and supporting commercial negotiations. The interpretation of large volumes of data from numerous sources requires a quick thinking and logical mind with the added ability to produce reports under pressure. Team players are essential, however you must also have the ability to work on your own initiative.

To be considered for some of the most exciting Analyst opportunities available, apply in the first instance by sending your curriculum vitae to Kerry Morrisoe quoting reference EA100, Capita RAS, Innovation Court, New Street, Basingstoke, Hampshire RG21 7JB, fax: 01256 383790, email: kmorr@rasnet.co.uk

Alternatively for further information please call William Sharp or Ari Richardson on 0171 202 0538.

An equal opportunity employer.

Energy.

It's our business.

CAPITA RAS EASTERN POWER & ENERGY TRADING

WANTED

The best and brightest Science and Engineering Graduates

The National Museum of Science & Industry is the world's pre-eminent museum devoted to the history and contemporary practice of science, medicine, technology and industry. The Museum is now poised for one of the most important developments in its 150 year history – a £45 million Wing to be devoted to contemporary science and technology with special reference to the new biosciences and information technology.

We need the brightest and most innovative young minds to help realise our vision. The Museum offers the best graduate development programme in our sector – and early career advancement for the most gifted.

If you are expecting a first or upper second this summer, or already have a good degree, preferably in engineering or a physical science, we should like to hear from you.

Starting salary is within the range £14,200-£15,500 depending on qualifications and experience. Other benefits include a non-contributory pension scheme, generous leave allowance and interest free season ticket loan.

Requests for further details please to Personnel Services, Science Museum, Exhibition Road, London SW7 2DD quoting ref. MT.

The closing date for applications is 17 July, 1998. Preliminary interviews will be held in the week beginning 27 July, 1998.

The National Museum of Science and Industry is an equal opportunities employer.

The National Museum of Science & Industry

Science Museum

BARNET HEALTH Authority

COMMUNICATIONS MANAGER

Salary: Circa £23K - Based North London

Barnet Health Authority commissions the best possible health care for the residents of the London Borough of Barnet. The health issues are challenging and we work in a high profile environment which calls for exceptional communication management to ensure that clear channels of communication are always in place. We have a vacancy for a full-time Communications Manager, for a fixed-term period of 2 years, working to the Director of Community Development and External Relations.

This job encompasses the full range of communication issues, including:

- **managing all media relations - the routine as well as the controversial;**
- **promoting the organisation's image through user friendly documents/information;**
- **developing and supporting internal communication;**
- **contributing to successful public involvement using innovative approaches.**

This job is not for the faint hearted! If you are looking for a job which will give you full exposure to the diversities of healthcare within a communications role, in a health authority that has a national as well as a local profile, and you have the drive and commitment to keep going when the going gets tough, this job is for you.

The skills and experience we are looking for are diverse so you need to be able to:

- **work under pressure to meet tight deadlines;**
- **come up with creative solutions whilst thinking on your feet;**
- **work proactively, continually looking for ways to develop and strengthen how we communicate internally and externally.**

The background and experience needed to do this job means that you are probably looking for at least your second career move. So, if you would like to find out more about the job and the organisation, call for an information pack on 0181 201 4703 or write to The Human Resources Department, Barnet Health Authority, Hyde House, The Hyde, Edgware Road, London NW9 6QQ, **quoting job reference CMI/00.** The closing date for receipt of applications is ***Friday 3rd July*** and interviews are expected to be held on ***Friday 10th July.***

Barnet Health Authority is working towards equal opportunities and welcomes applications from all sections of the community

SOFTWARE DEVELOPERS

Major investment banks, management consultancies and leading edge software companies are recruiting at a variety of levels. Skills required include Front, Middle and Back Office systems, Intranets and networking/communications/new media projects. Phd/Msc/Bsc's welcome.

Positions from graduate to senior project manager. TSC understand your career is important and pride ourselves on listening, and matching your aspirations as closely as possible, now or when a more relevant opportunity arises.

- **derivatives/equities**
- **intranets**
- **fixed income**
- **C/C++/SQL Server**
- **NT/Unix**
- **O-O design/analysis**
- **Orbix/Corba/COM**
- **ORACLE/SYBASE**
- **futures/options**
- **visual basic/java**
- **quantitative analysis**
- **client/server**

the soft corporation

recruitment specialists in financial markets

the soft corporation
7-15 Rosebery Avenue, London EC1R 4RP
Tel: 0171 833 2772 • Fax: 0171 833 2774 • Email: jmcb@softcorp.demon.co.uk

FOCUS Job adverts and descriptions

1 TASK

Before you read the job advertisement for engineers in section 3, discuss the following questions in small groups.

1 Which of the following information do you expect the text to provide:
- details about the salary?
- information about the company?
- examples of typical products of the company?
- the experience the company expects people to have?
- a clear job description?
- where the company is located?
- whether the company provides training?

2 What other information do you also expect to find?

3 In which order do you expect the information to be given?

- Compare your results, then do section 2 before you read the advertisement.

2 PRACTICE

Phrases 1 – 7 are from the advertisement. Expressions a – g mean the same as the underlined parts of 1 – 7. Match the expressions.

1 ... the career prospects and rewards package ...
2 ... leadership does not come out of the blue ...
3 ... to assume a prominent role in their countries ...
4 ... communication lies at the heart of our business ...
5 ... our leading-edge products ...
6 ... our commitment to excellent service ...
7 ... technical and personal development is an integral part of a career with ...

a ... just appear ...
b ... is an essential element of ...
c ... highly advanced ...
d ... strong belief in the importance of ...
e ... take a leading position ...
f ... the chances of promotion and the salary and other benefits ...
g ... the initial and ongoing training that we provide ...

- Which areas of information that you identified in section 1 would you expect to find phrases 1 – 7 in?

3 A job advertisement for engineers

READING

Read the advertisement and compare your results from sections 1 and 2 with the text.

Can world communi-cations leadership come out of the blue?

Telecom 3000 exists in order to help the world communicate more easily. Whether you use land lines or GSM technology, whether you send data to the other side of the country or speak to someone on the other side of the world, or simply use a mobile phone to order a take-away pizza, the chances are high that you use our technology when you do so. Communication lies at the heart of our business.

Technology like this does not come out of the blue. Our leading-edge products are the result of years of hard work, developed in close co-operation with our customers. Although what we have produced answers many general needs, our products are the result of responding to the specific needs of our customers. Now more than ever, we are turning towards our customers as long-term business partners, and the people we want to appoint to our Public Switching Division must share our commitment to excellent service.

Specifically we are looking for the following

- **Technical Support Engineers**
- **Customer Support Engineers**
- **Systems Engineers**

For all these areas we have opportunities to accommodate a range of experience – from graduates with only limited support or development experience through to more senior specialists with between two and five years'practical background who want to develop their careers in a team leadership/management role. Telecom 3000 is preparing a stronger presence on the Continent and thus also welcomes applications from European citizens, in particular from France, Germany and Spain, who would be willing to assume a prominent role in their respective countries after a few years in England.

Detailed and ongoing technical and personal development is an integral part of a career with Telecom 3000. However, as a foundation, a degree/HND in Computing/Communications, or equivalent telecommunications experience ideally with a knowledge of Telecom 3000 systems, for example DMS, would be highly valued.

The career prospects in this rapidly changing, innovative company are excellent. The rewards package is comparable with the best the industry has to offer.

To apply, please fax a typed CV, including salary details to our Recruitment Centre on 0181 945 3018 clearly quoting reference ENA00105. Alternatively send your CV, quoting the above reference to Philip Worthing at our Recruitment Centre, Telecom 3000 Europe Ltd, Headquarters Building, Salford Road, Manchester MC5 1AY.

Telecom 3000

4
LISTENING

After seeing the Telecom 3000 advertisement, Carol Mistadou phones Mick Henderson. Listen to the first part of their conversation and answer the following questions:

1 What do you think the relationship between Carol and Mick is?
2 Why is she phoning?
3 What has changed in Carol's and Mick's lives since they last spoke?

- In the second part of the conversation, Mick gives job descriptions for systems engineers, technical support engineers and customer support engineers.

Copy the table, then listen and fill it in.

	Customer contact (yes/no)	Main responsibilities
Systems engineers		
Technical support engineers		
Customer support engineers		

What kind of job do you think Mick has? What does he say about his holiday plans?

GRAMMAR

Simple past – past continuous

1	I **was changing** the nappies when the phone rang.
2	You **were not doing** too well the last time we spoke.
3	I **was wondering** if you could give me a lift?

- The past continuous is used for activities in the past which continued for some time and were in progress at a given time. (sentences 1 – 2)
- The past continuous is also used to ask questions in an indirect way. (sentence 3)

5
PRACTICE

Fill in the gaps with the past simple or past continuous.

1 … *(I/get/frustrated)* when I suddenly found the solution.
2 I … *(already/plan/change/jobs)* when I saw this advertisement in the paper.
3 The last time … *(I/see/her/she/live/in SF)* , but I think she has moved in the meantime.
4 What was I doing yesterday at 2 o'clock? Let me think. Yes, I know, … *(I/have/lunch/with/David)*.
5 I … *(see)* Sarah yesterday; she … *(stand)* practically in the middle of the road talking to someone on her cellular phone.
6 Tom … *(waiting)* for me when I … *(arrive)*.

FOCUS An application for a job

6 Carol Mistadou applies for a job

READING

Carol wrote the following letter of application.

Dear Mr Worthing

Ref: ENA00105

With reference to your advertisement in The Times of October 19th, I should like to apply for a position as a Technical Support Engineer.

I am a French national and have just completed my studies in Telecommunications Engineering at the Technical University of Toulouse. During the final years of my studies, I took a special interest in the interface of design and function aspects of modern consumer electronics, and participated in a research project on making multipurpose telephones more consumer-friendly. As you will see from my CV, I attended a "Lycée Technique" and held a position as a research assistant prior to my studies in Toulouse.

I am very interested in the position because it would seem to offer the opportunity to further my knowledge and pursue a career in a field that I feel I am very suited for. In addition, I have always wanted to work in England for a few years and would later welcome the challenge of a responsible position in France. Having spent four years with an international company in which the working language in the research laboratory was English, I expect no language problems.

Enclosed you will find my CV, photocopies of my diplomas, as well as letters of recommendation from both my professor and my former employer. I would appreciate the opportunity to discuss this application at any time.

Yours sincerely

Carol Mistadou

PHRASES

Letters of application

Introductions:

I should like to apply for …

I wish to be considered for …

I am interested in the position advertised in …

With reference to …

In reply to …

Employment/Qualifications:

I attended …

I served an apprenticeship as a mechanic with …

I graduated from …

My main subjects …

I specialized in …

I have been employed by …

I have been working for …

Reason for the application:

It would seem to offer better opportunities to …

I would welcome the challenge of …

I am keen to be employed abroad …

Conclusion:

I would welcome the chance to …

I should appreciate the opportunity to …

I am available for an interview …

Yours sincerely

7
TASK

Telecom 3000 has also received an application from a German student whose CV you see below. Write a letter of application to Telecom 3000 based on this CV. When you have finished you may compare "your" application with the example given on page 134.

CURRICULUM VITAE

Surname:	Rüteling
First names:	Christian, Thomas
Address:	Inselstrasse 32, 69418 Frankfurt, Germany
Date and place of birth:	14.9.1967, Aachen, Germany
Nationality:	German
Marital Status:	Single

Education

1977 - 1983:	Gesamtschule Hehrenberg, Aachen (Comprehensive School)
1983 - 1987:	Lehre als Kommunikationselektroniker (Apprenticeship in telecommunications electronics)
1991 - 1992:	Fachoberschule, Aachen (One-year preparatory course for university)
1992 - 1994:	Fachhochschule Aachen: Diplomstudiengang Elektrotechnik (University of Applied Science: Electrical engineering)
1994/95:	King's College, Plymouth Exchange student
1995 - 1997:	Fachhochschule Dieburg

Qualifications

12.7.1983:	Zeugnis der mittleren Reife (Comprehensive School: Diploma of General Secondary Education)
6.6.1987:	Gesellenbrief (City and Guilds Certificate in telecommunications electronics)
16.6.1992:	Fachschulreifezeugnis (Diploma of General Secondary Education, university entrance qualification)
17.6.1997:	Diplom in Kommunikationselektronik (University of Applied Science diploma in telecommunications engineering)

Employment

1983 - 1987:	Deutsche Bundespost, Aachen
1987 - 1992:	Deutsche Bundespost, Köln
1992 - 1997:	Datacom (part-time)
Languages:	English (fluent) French (good)
Leisure Interests:	Music, travel, sport

FOCUS Booking a hotel

8 LISTENING

Carol Mistadou books a room

Carol Mistadou has been invited for a job interview in England. To book a hotel she makes a few phone calls. Listen and answer the questions below.

1 What dates does Carol want to make a reservation for?

2 What kind of rooms are available at the Heartwood Hotel and how much do they cost?

3 What kind of rooms will be available at what time and for what price at the Green Leaves Hotel?

4 What room does Carol book?

5 What does she have to do after the phone conversation?

PHRASES

Booking a hotel

I'm calling to inquire about ...

I'd like to inquire about ...

I'd like to make a reservation for ...

How much do you charge for ...?

Does that include breakfast?

Are there any vacancies ...?

Is that en suite?

Do you accept credit cards ...?

When will you be arriving?

How long are you staying?

Are you paying by credit card?

That's £55 for a single and £75 for a double.

Could you confirm the reservation in writing, please?

9 TASK

Work in pairs. In Pairwork Task 8, Partner A wants to book hotel rooms; Partner B is the hotel receptionist.

A: Go to Pairwork Task 8, page 130. B: Go to Pairwork Task 8, page 133.

Focus Mobile telephone systems

10 READING Inside the cellular system

Mick Henderson has found an article about mobile telephone systems in a daily paper. He thinks that it may help his friend Carol prepare herself (language-wise) for the interview. He therefore sends her the article. He has marked a number of words which he knows are difficult to pronounce for French speakers of English.

When you read the text, pay special attention to these words.

TECHNOLOGY

INSIDE THE CELLULAR TELEPHONE

A look inside cellular telephones, and the fascinating technology that has revolutionized mobile communications.

JOSEF BERNHARD

PRIOR TO THE DEVELOPMENT OF the cellular system, mobile telephone systems relied on centrally located transmitting and switching equipment to communicate with mobile units. Each transmitter had a large area of coverage and only a few communication channels were available within this area. Cellular systems, on the other hand, divide their region of coverage into many small areas, each of which is only a few square miles in size. It is this territorial subdivision that allows the mobile units to use low-powered transmitters, and to use and reuse the same frequencies in the same system to increase the number of communication channels available.

Close to a thousand two-way communication channels are allocated for cellular service but not all of them can be used for communication between subscribers to the system. Approximately 50 channels are needed for control signals which coordinate the frequencies and assure smooth operation of the system.

When a mobile unit leaves the region of coverage of one cell site and enters that of the adjoining one, it is said to be "handed-off" from one cell to the other, and a lot of things happen simultaneously during this fraction of a second. As a mobile phone traverses a particular cell, it eventually reaches a point where the signal is no longer strong enough for reliable communications. Fortunately, by the time it has reached that point, it is well within the region of coverage of an adjacent cell. The handing-off process of transferring the responsibility for a call-in-progress from one cell site to another requires a lot of "intelligence" on the part of both the cell and the mobile unit.

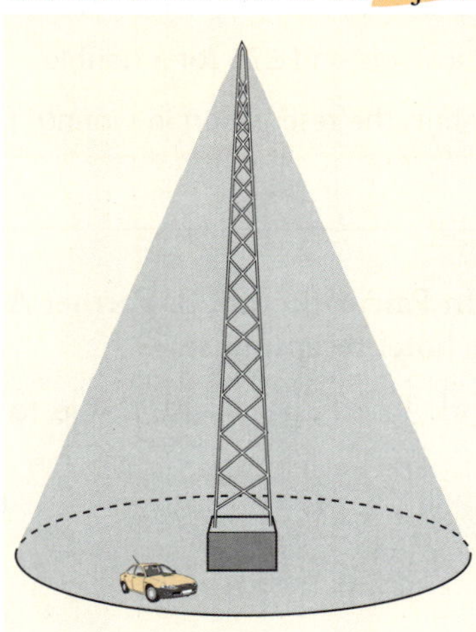
Early mobile telephone

The first thing that has to be done is to sense when a signal is approaching the point where it is about to become weak. That's easy; all that is needed is a signal strength detector. More complicated is the task of determining which cell site is to receive the hand-off. The cell sites in the system have to "confer" to see which of them is receiving the signal best, and make arrangements for transferring the call without interruption. This is not too difficult either. The next part, however is more complicated.

Because adjacent cell sites cannot use the same frequencies, even though others in the same system can, a new set of frequencies must be used after the hand-off. And since the new cell site will be using different frequencies, so must the mobile unit. That is where the control channels, and the intelligence built into them, come into play. The new cell site knows which of its frequencies are in use and which are free, and makes a decision to allocate one of its frequency pairs (one channel) to the new conversation entering the district. It then transmits, over the control channel, instructions for the cellular phone to switch to those new frequencies. The phone adjusts the voltage of its frequency synthesizer accordingly, and the conversation continues in the new cell site on the new frequencies – all of that with only an unnoticeable interruption of a millisecond or two.

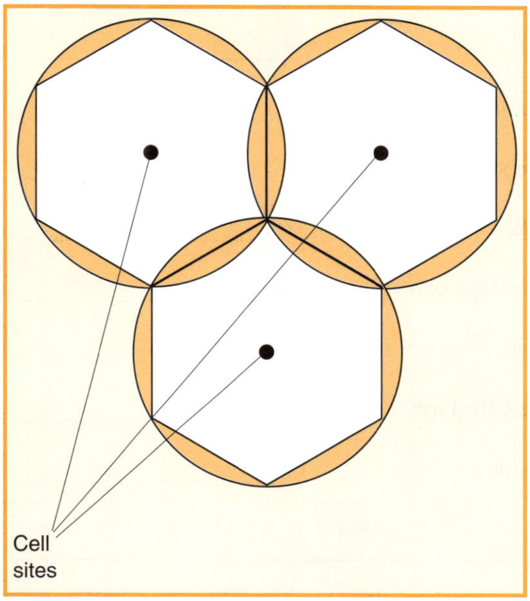

Cellular system

Another interesting feature of cellular phone systems is the constant monitoring and adjustment of power levels. One of the rules of cellular telephony is to use only as much power as needed. Consequently, a cell site monitors the strength of the signal it receives from a mobile unit. If the strength increases to a predetermined level, the cell site sends instructions over the control channel for the mobile unit to reduce its power to a lower level. Conversely, if the received signal drops, a mobile phone can be instructed to increase its power. Cellular phones are capable of between 3 and 8 discrete output levels. Keeping output power to the minimum required for good communications also reduces the risk of interfering with communication in a nearby cell.

11 TASK

Discuss which of the following three short summaries of the text in section 10 is the best.

1. In the cellular mobile telephone system many small areas of coverage work together to create a very large system in which more than 2000 two-way communications are possible at the same time. Signal strength detectors constantly monitor the strength of the signals and if a signal becomes weak, the call is transferred to another cell. To avoid interference between the cells, the power levels of the mobile unit are kept at a minimum.
2. Cellular mobile telephone systems use intelligent monitoring and adjusting techniques which allow close to a thousand calls to be made simultaneously in each cell. Each call uses one frequency, but the frequency is changed if the caller leaves a cell and enters a new cell. 50 of the two thousand frequencies are reserved for monitoring and adjusting purposes.
3. The advantages of a cellular mobile telephone system are that more channels are available and less energy is used than in conventional mobile systems. A cellular system uses intelligent electronics to monitor frequencies and energy use, adjusting them as necessary.

12 PRACTICE

Copy and complete the table below by filling in words from the text in the middle column.

synonym	word from the text	opposite meaning
before	1 ________	after
to lower	2 ________	to increase
to depend on	3 ________	to be independent of
about	4 ________	precisely
to go away	5 ________	to enter
at the same time	6 ________	later/before/another time
to take	7 ________	to give
to check/to control	8 ________	to disregard
next to	9 ________	not close to/away from
to pass through	10 ________	to remain within
movable	11 ________	immobile

CULTURE

What is important to get a job[1]

	Personal connections not important = 1 very important = 5	Formal qualifications not important = 1 very important = 5	Three most important character traits
Germany	2	4	efficient, serious, reliable
Britain	2	2 – 5[3]	efficient, managing skills, good communicator
USA	3	2 – 3	serious, efficient, honest
Japan	2	2 – 3[4]	harmonious, group mentality, efficient
Scandinavia	1	2	efficient, relaxed, serious
Spain	2 – 4[2]	3	charming, efficient, good communicator
France	4	4	efficient, reliable, flexible
Italy	2 – 3[2]	4	efficient, reliable, serious

[1] The table shows basic trends; behaviour may vary.
[2] more important in small companies
[3] very important at start of career, less important later
[4] name of university is very important

Modules

Manufacturing

MODULE 1a

Automation

Focus: Drawing and turning
Focus: Sensors in manufacturing

MODULE 1b

An assembly plant

Focus: Authentic reading
Focus: Using a dictionary

Energy

MODULE 2a

Solar energy

Focus: Solar heating
Focus: Solar intensity

MODULE 2b

Atomic power

Focus: Authentic reading
Focus: Learning vocabulary

Electricity

MODULE 3a

An electric motor

Focus: Magnetism and electricity
Focus: An AC alternator

MODULE 3b

Instructions

Focus: Authentic reading
Focus: Tests

Architecture

MODULE 4a

Structures

Focus: Basic structures
Focus: The history of bridges

MODULE 4b

A television tower

Focus: Illustrations in technical texts
Focus: Authentic reading
Focus: Writing skills

Recycling

MODULE 5a

Providing and recycling

Focus: Water treatment
Focus: Sorting techniques in recycling
Focus: Waste incineration
Focus: Recycling plastic

MODULE 5b

The tyre problem

Focus: Authentic reading
Focus: Giving a presentation

Tele-communications

MODULE 6a

Waves and systems

Focus: Waves
Focus: A basic telecommunication system

MODULE 6b

Telecommunication links

Focus: Authentic reading
Focus: Features of technical texts

MODULE 1a

Automation

Focus Drawing and turning

1 VOCABULARY

Crankshafts and camshafts can be used to convert rotary motion into special types of linear motion. In small groups study the illustrations and answer the questions that follow. If necessary consult your dictionaries.

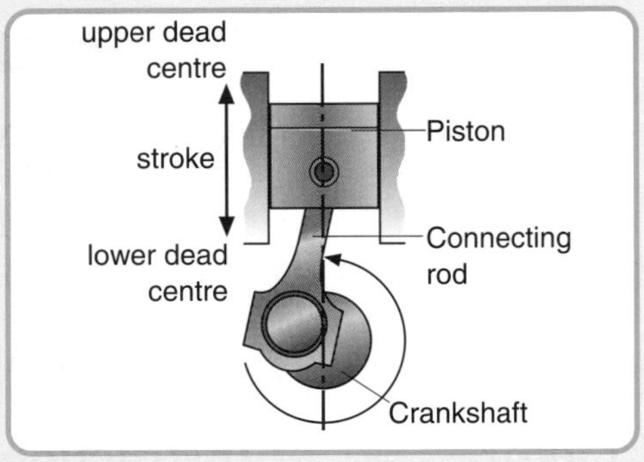

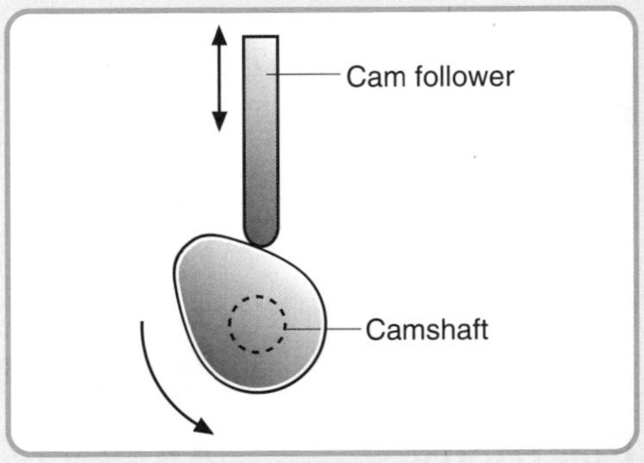

1 What is the meaning of "upper dead centre", "lower dead centre" and "stroke"?
2 What is the main difference in the kind of motion that a camshaft and a crankshaft produce?
3 Name three examples of the use of crankshafts or camshafts.
4 What other mechanisms can you think of that convert motion?

2 READING

Early production machinery

Figure 1 shows the process of "wire drawing". In wire drawing, a thin wire is produced by pulling a thick wire through a small hole (the die) to reduce its diameter. *Figure 2* on the next page shows a lathe. Lathes are used, for example, to reduce the diameter of a round workpiece.

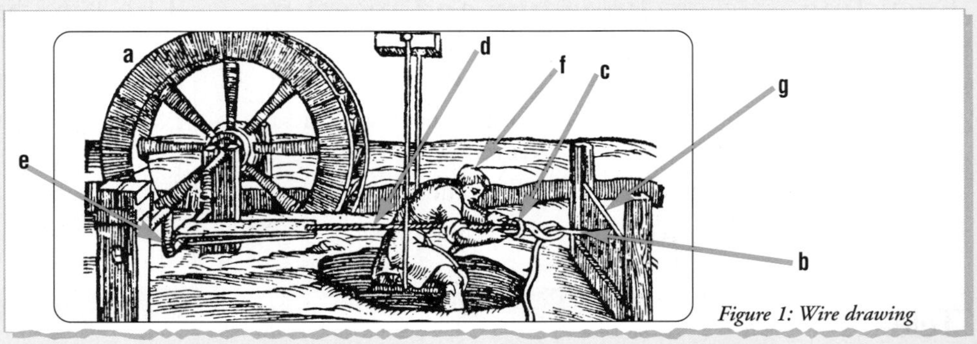

Figure 1: Wire drawing

Before you read the texts, look at the illustrations and try to identify some of the parts in the following list. Then read the texts and – if necessary – complete the task.

Wire drawing: water wheel, wire-drawing bar, rope, operator, crankshaft, wire, pair of pliers; **Lathe:** flywheel, fixed centre, foot treadle, running centre, crankshaft

Mechanization of production processes began very early, much earlier in fact than the time we usually call the Industrial Revolution, when steam began to be used as a source of power. For centuries, water mills had provided the power for many industrial processes such as the process of wire drawing shown in *Figure 1*. The illustration shows a medieval water wheel with a crankshaft attached to its centre. Wheel and crankshaft rotate continuously. The two ends of the crankshaft are mounted in bearings, and the crankshaft is attached to a pulling device or rope. The other end of this rope is fastened to a pair of pliers in such a manner that, when the rope is pulled, the jaws of the pliers close firmly around the wire. The operator grips the wire when the position of the crank is at the dead centre closest to the wire-drawing bar. As the crank continues its rotation, the wire is pulled through the die in the wire-drawing bar. This lengthens the wire while reducing its diameter. At the opposite dead centre, the operator releases the pliers, and during the following forward stroke the pliers slide back until the crank is back at the starting point. He now grips the wire again with the pliers, and the cycle is repeated. In such a manner a great length of wire can be drawn; the operator needs to apply only little force, as all the pulling force is provided by the water wheel. If a further reduction of the wire diameter is necessary, a smaller diameter die is used, and the process is repeated. When the wire is first threaded through the die, the tip of the wire must be sharpened to allow it to be pushed through the die far enough for the operator to grip it with the pliers. Only soft materials such as copper could be drawn in such a manner. To reduce friction, the die also needed frequent lubrication.

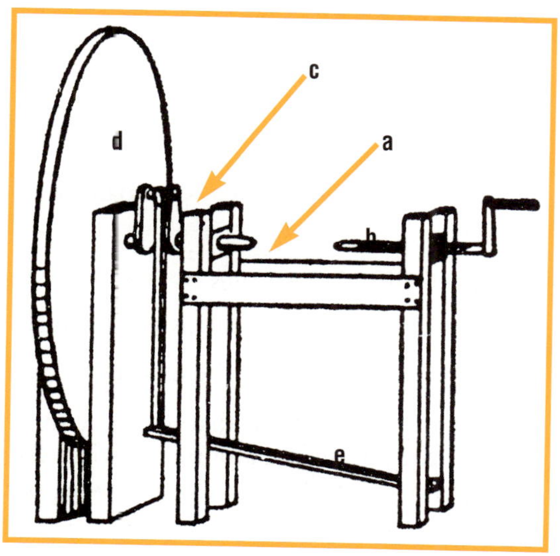

Figure 2: Leonardo da Vinci's Lathe

Figure 2 shows a sketch made by Leonardo da Vinci: the design of a lathe for the turning of wood. An interesting feature of this design is the flywheel attached to one end of the crankshaft. The crankshaft is rotated by the up-and-down movement of the foot treadle. Due to the weight of the flywheel, the motion of the crankshaft is made more uniform. The running centre of the lathe is attached to the opposite end of the crankshaft, and the workpiece to be turned is mounted between the running centre and the fixed centre. To fit different lengths of workpieces, the distance between the two centres can be adjusted.

3 PRACTICE

In technical texts the passive is often preferred when processes are being described. Read the text again and write down all the passive constructions. Why is the passive used?

(For an explanation of the passive go to Unit 1, page 14)

- Now make passive sentences as in the example. Remember to add the correct form of the auxiliary verb 'be'.

1 The ends of the crankshaft	convert	to a pulling device.
2 The wire	mount	to thread it through the die.
3 The oscillating movement of the foot treadle	must/sharpen	when the wire is drawn through the die.
4 The tip of the wire	pull	through the die by means of a pair of pliers.
5 The diameter of the wire	achieve	by means of a flywheel.
6 The crank	reduce	in bearings to reduce friction.
7 Uniform rotation of the workpiece	attach	into rotary motion of the workpiece.

4 VOCABULARY

The verbs *allow, cause, permit, enable* are similar in meaning. They all express ability or a cause-and-effect relationship between two things. *Enable* is usually only used with a human object.
Reformulate the sentences below using the words in brackets.

1 Rotary motion is transformed into reciprocating motion by means of the crank. *(permit/transformation of)*

2 The operator is positioned flexibly in the basket and therefore can follow the lateral movement of the rope. *(Flexible positioning/enable)*

3 The operator has to apply only little force because the pliers are self-enforcing. *(self-enforcing mechanism/allow)*

4 The flywheel ensures that the lathe rotates smoothly. *(cause)*

5 DISCUSSION

In groups of three or four discuss the following questions. Compare your results with the other groups.

1 Was it necessary to have the operator sitting in a swing suspended from above?

2 What was the shape of the dies in the wire-drawing bar? Make a sketch.

3 Did it matter whether the water wheel rotated clockwise or anticlockwise?

4 The pliers are of the self-enforcing type. This means that the more force is applied to pull the rope, the firmer the pliers close around the wire. What was the shape of the pliers to make them self-enforcing? Make a drawing of the pair of pliers.

5 The wire drawing illustrated in Illustration 1 shows an intermittent process, which means that the wire is not stretched continuously but bit by bit. How do you think a continuous drawing process could have been achieved?

FOCUS Sensors in manufacturing

6 Types of sensors

READING

The following short text classifies sensors according to output and according to function. Decide what categories of sensors you expect it to name, then read and check.

TYPES OF SENSORS

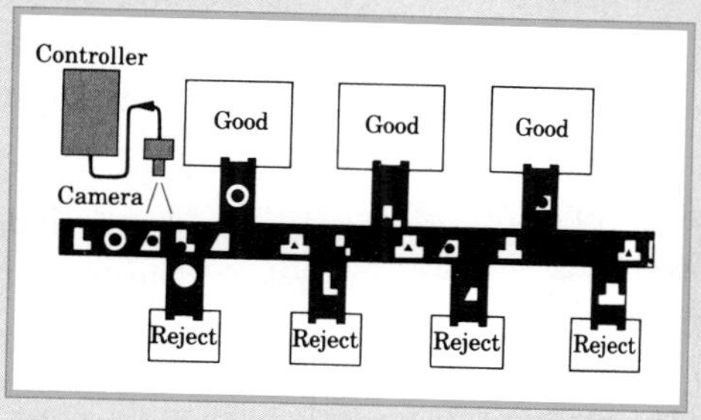

Most modern manufacturing techniques would not be possible without the use of sensors. A sensor is a device that produces a signal to detect or measure a property such as, for example, the position or temperature of a workpiece, or the force being applied to it. Traditionally, sensors are used to set limits on the performance of machines. Examples are the stops on machine tools which restrict the movement of the worktable.

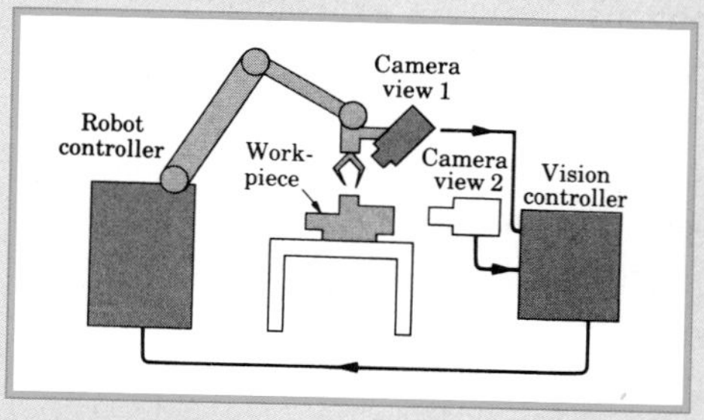

There are two different types of sensors. Analogue sensors produce a signal that is proportional to the measured quantity, whereas digital sensors have numeric or digital outputs that are transferred directly to computers.

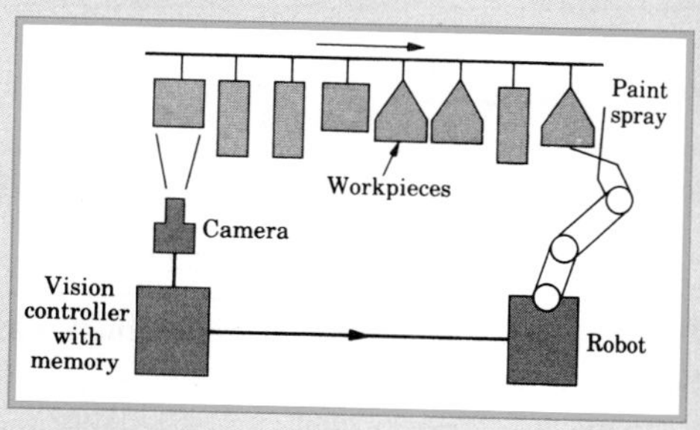

Sensors are frequently found in assembly operations. They are classified as either tactile sensors or visual sensors. Systems using tactile sensors are used to handle fragile parts such as glass bottles or electronic devices. In visual sensing, cameras optically sense the presence and shape of an object.

102

7 PRACTICE

Now read the rest of the text about sensors and decide whether the verbs in brackets should be in the passive or active. Sometimes you may want to add a modal verb (*can/must/should/may* etc).

TYPES OF SENSORS

In applications where fragile parts are handled, the robot's gripper ...[1] *(equip)* with a sensor such as a piezo electric device or a strain gauge that is capable of measuring the force being applied. The force sensed can be monitored and controlled to such an extent that even slipping ...[2] *(sense)* and thus prevented, a capability of human fingers and hands that we ...[3] *(take)* for granted but which is very difficult to achieve in robotics.

Advanced visual sensing systems can sense up to three dimensions and are, for example, capable of detecting properly inserted components in a printed circuit or a properly made solder joint. When they ...[4] *(use)* in automated inspection systems, they also ...[5] *(detect)* cracks and imperfections.

In visual sensing, the visual image of the object ...[6] *(measure)* by a microprocessor, the measurements ...[7] *(digitize and compare)* with an equally digitized image of the ideal shape or orientation of the object.

Machines equipped with visual sensors can identify and inspect parts and reject defective ones. Grippers are able to pick up parts and put them into the correct position. However, picking parts from a bin is a difficult task, because of the way they are randomly arranged very close to each other.

GRAMMAR

Reduced relative clauses I

1. Illustration 2 shows a sketch **which was made** by Leonardo da Vinci.

   Illustration 2 shows a sketch **made** by Leonardo da Vinci.

2. An interesting feature of the design is the flywheel **which was attached** to one end of the crankshaft.

   An interesting feature of the design is the flywheel **attached** to one end of the crankshaft.

Relative clauses with a passive are often reduced by leaving out the relative pronoun and the auxiliary *be*, leaving just the past participle.

- Now read the texts in sections 6 and 7 again. How many reduced relative clauses can you find?

8 VOCABULARY

Some verbs can be changed into nouns by adding *-ment* or *-ion*, or *-ity/-ivity*, for example *equip/equipment, automate/automation, produce/productivity*. Other verbs do not change, e.g. *to change/the change*.

What are the nouns that belong to the following verbs? Use your dictionary to check spelling changes.

1 operate	________	12 require	________
2 repeat	________	13 measure	________
3 pull	________	14 be able	________
4 grip	________	15 produce	________
5 restrict	________	16 describe	________
6 lubricate	________	17 push	________
7 reject	________	18 transfer	________
8 adjust	________	19 prevent	________
9 attach	________	20 rotate	________
10 achieve	________	21 apply	________
11 illustrate	________		

PHRASES

Describing a function or purpose

The lathe **serves to** reduce the diameter of a round workpiece.
The function of the lathe **is to** reduce the diameter of a round workpiece.
A lathe **can be used to** reduce the diameter of a round workpiece.
The purpose of a lathe **is to** reduce the diameter of a round workpiece.
The diameter of a round workpiece can be reduced **by means of** a lathe.
A lathe **is for** reducing the diameter of a round workpiece.

- Express the function or purpose of the following:

1 tactile sensors
2 visual sensors
3 the stops on machine tools
4 strain gauges
5 grippers
6 da Vinci's flywheel

9 Understanding servo-mechanisms

LISTENING

The illustrations below show different kinds of servo-mechanism that are described in a lecture on the cassette.

First study the diagrams. Several items are missing. As you listen, note down the missing items and where they should be, and the names of the parts indicated by arrows and numbers. Make sure you know the meaning of the words 'amplify' and 'amplification' before you start.

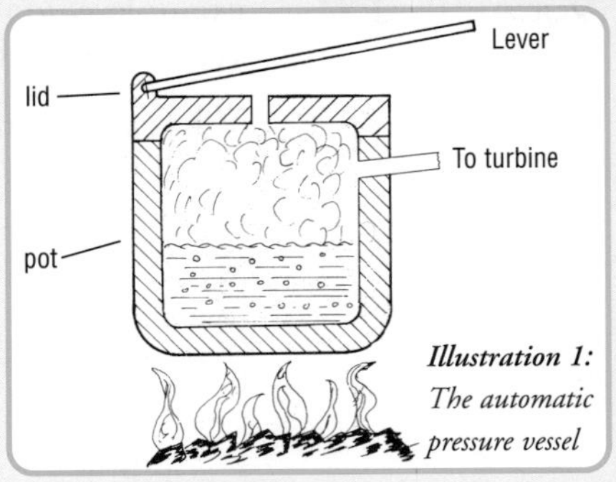

***Illustration 1:** The automatic pressure vessel*

***Illustration 2:** The transistor*

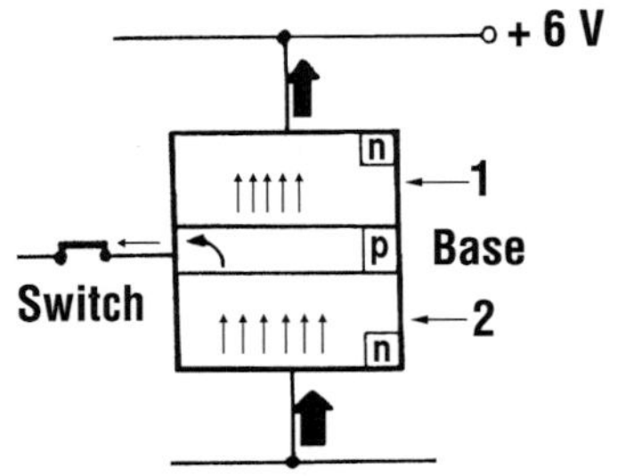

***Illustration 3:** A hydraulic clamping device*

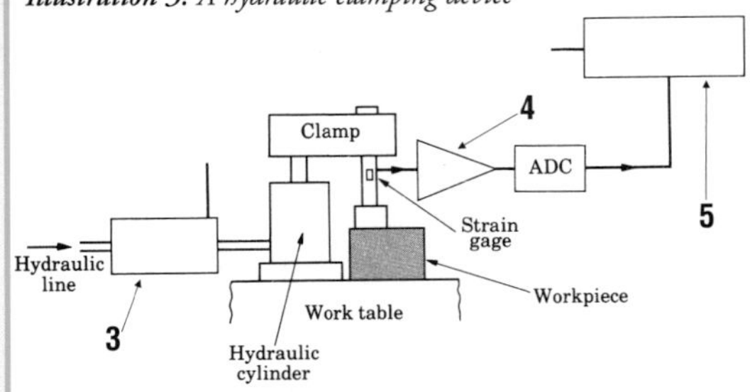

10

WRITING

Write a description of the plastic extruding process shown in the illustrations below. The following information will help you.

Purpose Plastic Granules → pipes/rods/films etc.

Process Granules → Hopper → Feed zone (granules are mixed) → Melting zone (granules begin to melt) → Melt-pumping zone (melting and increasing of pressure) → Die (shape given) → Cooling of material

Details
- Diameter of the screw increases to increase the pressure
- Continuous process (great length can be produced)
- Great variety of shapes can be produced

***Illustration 1:** Plastic extruding*

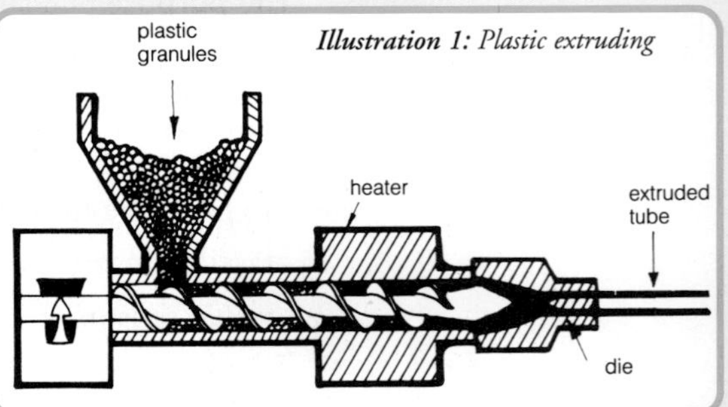

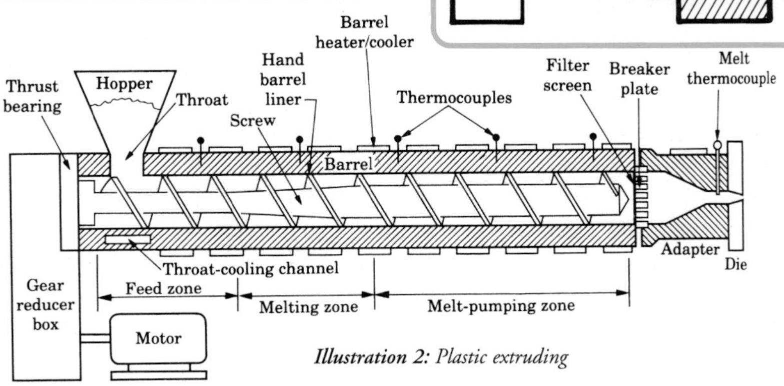

***Illustration 2:** Plastic extruding*

MODULE 1b

An assembly plant

Focus Authentic reading

1 Toyota sacks some robots

READING

First look at the heading and the subheading of the following article. What do you think the text is going to be about?

- Now read the text and decide which two of these statements are correct:

1. Toyota is reducing automation in all of its plants.
2. Toyota has discovered that it is more economic to have fewer robots and more humans than a few years ago.
3. Toyota is replacing robots by humans because they are cheaper.
4. Toyota has redesigned its assembly lines and increased the number of subassembly lines.

Toyota Sacks Some Robots

Manufacturing:

Japan's famously automated automakers emphasize the human touch at home

TOYOTA'S NEWEST CAR FACTORY is an industrial wonderland. Sparks fly as floors, side panels and roofs are welded. Robots install parts, while humans thread electrical wire and put in brake linings. Overhead, an electronic board tracks productivity; at midmorning, the plant has produced 145 cars - two behind its 480-cars-a-day schedule. "We're probably behind because we just came back from a long holiday," says a guide. "We're usually running at 105 percent."

Toyota's $1,4 billion Kyushu factory is arguably Japan's most modern – but not in the ways one may expect. The country has long been at the forefront of factory automation and robotics, but in designing the Kyushu plant, Toyota chose to rely on less automation, not more. Jobs performed by robots at other Toyota car factories have been given back to humans. "More automation is not necessarily our aim," says Tadaaki Jagawa, an adviser to the Kyushu plant. "We want manufacturing that is done with human skill and wisdom."

The main reason Japan's leading carmaker has slowed its march toward automation at its Japanese plants is that it doesn't think that course will yield any further significant cost savings. In the past, adding machines meant subtracting workers. But having ridden a wave of automation in the 1980s, Japanese carmakers have reduced labor to only 6 percent of total production costs. There is just not that much more left to save. But Toyota also learned an unexpected lesson after spending heavily in 1991 to set up a new assembly plant at its massive Thara complex. Some of the state-of-the-art machines it installed proved too complex and unreliable. Assembly workers were demoralized by sophisticated equipment they could neither operate nor maintain effectively. As a result, Jagawa says, the plant struggles financially.

Tadayuki Kanedo, a general manager of Tokyo-based Toyo Engineering Corp., says that in the rush to automate, many Japanese manufacturers made a mistake. They took the most interesting assembly work away from humans and gave it to robots, which must be programmed every time there is a model change. With frequent model changes, retooling and engineering costs climb. "It's not good to introduce technologies that threaten the workers at a plant. We'd like to manufacture cars in a way that maintains communication between autoworkers and customers." As a practical matter, Kanedo says that a plant's automation level is not a good measure of its efficiency. A factory with a 30 percent automation rate and 250 workers may be less cost-effective than one with 400 workers and a 10 percent automation rate.

Toyota Kyushu's most innovative concept may be its redesigned assembly line. Most automobile plants have three to four subassembly lines, with roughly 70 workers on each. If something goes wrong, an entire line usually stops. Kyushu has broken the big lines into 11 mini-lines, each tended by a kumi, a group of 15 to 20 workers. Each kumi, in turn, is segmented into three teams, or han, that handle specific tasks such as the installation of steering wheels. Each kumi is responsible for the quality of work on its line. If there is a problem, fewer workers are left standing around, and the other 10 lines can keep moving. "The mini-lines have improved morale significantly, and because of that, productivity has also improved."

Though Toyota is re-emphasizing the human touch, the company realizes there is no getting away from advanced technology. Computer-controlled machines not only do the heavy work – lifting body panels and engines – but also perform such precision tasks as measuring engine blocks by microns and conducting all sorts of diagnostic tests. Humans could do that work, but not as well or as cheaply. "Toyota has not soured on automation," Jagawa says, "it has just gotten more selective about how it is used. Without human skills, technical advancement will fail. *(Newsweek)*

- Now answer the following questions on the text.

5 In what respect does the new plant not meet visitors' expectations?

6 What development marked the manufacturing industry in the 1980s?

7 What percentage of total production costs are represented by labour costs?

8 The text says that in highly automated factories the workers may become demoralized. Explain why.

9 In what respect does Toyota's concept of car assembly at the Kyushu plant differ from other car plants?

2 VOCABULARY

Use words from the text to replace the underlined words in the following sentences. (The figures in brackets indicate the lines of the text in which the word is used.)

1 Robots can <u>do</u> jobs that are difficult for humans to do. (1 – 34)

2 Some <u>very modern</u> machines have proved to be unreliable. (35 – 57)

3 The Japanese learned that the new machines were expensive to <u>service</u>. (57 – 63)

4 The company's most <u>creative</u> aspect may be its division of labour.(64 – 95)

5 The workforce is <u>grouped into</u> several teams. (96 – 117)

6 Toyota's car factory continues to rely on <u>sophisticated</u> technology. (117 – 129)

7 Many human <u>capabilities</u> are not matched by robots. (130 – 137)

3 SUMMARY

Prepare a written or oral summary of the text concentrating on the most important aspects.

4 DISCUSSION

In the past few years more and more people have been made redundant, partially due to new technologies in manufacturing. In many European countries 10% and more of the workforce are unemployed. Could a return to "old production techniques" help, or do we have to hope for "new markets" and "new" products or services in, for example, telecommunications?

FOCUS Using a dictionary

5
STUDY SKILL

Study the dictionary entry for the word 'manufacture'.

1 Read out the phonetic transcription. The ˈ tells you that the following syllable carries the main stress. Therefore it is manu<u>fac</u>ture.

2 A lot of words can have several meanings. The dictionary marks the different meanings with numbers.

3 (v) and (n) tell you whether the word is a verb or a noun.

manu·ally /-jʊəlɪ/ *adv*: *manually operated*.
man·u·fac·ture /ˌmænjʊˈfæktʃə(r)/ *v* [Tn] **1** make (goods) on a large scale using machinery: *manufacture shoes, cement, cookers* ○ *manufacturing industry*, eg in contrast with industries which do not make products. **2** (*usu derog*) invent (evidence, an excuse, etc): *She manufactured a false story to hide the facts.*
▷ **man·u·fac·ture** *n* **1** [U] activity of manufacturing: *firms engaged in the manufacture of plastics* ○ *goods of foreign manufacture*, ie made abroad. **2 manufactures** [pl] manufactured goods or articles.
man·u·fac·turer *n* person or firm that manufactures things: *Send these faulty goods back to the manufacturer.* ○ *a clothing, a car, an electronics, etc manufacturer.*
ma·nu·mi... mænjʊˈmɪt/ *v* (-tt-) (*fml*) (formerly)
...ɪʃn/ *n*

4 The symbol (Tn) tells you that the verb is transitive, i.e. it must be followed by an object. So you cannot just say "I manufacture", you have to say what it is that you manufacture: "I manufacture cars/bikes/shoes etc." The symbol (I) would mean that the verb was intransitive and was used without an object.

5 (usu derog) indicates that this particular use of manufacture has a negative meaning. (derogatory = showing a critical attitude)

6 [U] indicates that this noun is not countable.

- Now check your dictionary for the meaning and pronunciation of the following words:

1	technology	4	maintain	7	significantly
2	efficiency	5	sophisticated	8	productivity
3	demoralize	6	automation	9	perform

MODULE 2a

Solar energy

Focus Solar heating

1 Water circulation systems

READING

The following text is taken from a leaflet giving an introduction to solar heating systems. Read the text and identify items labelled A – G in the two diagrams.

Probably the most common solar energy technology in use today is the one employed in heating potable water. Over the past 15 to 20 years several different systems have been developed and constantly improved so that it is now safe to say that most homeowners will have no problems finding the right solution for their specific needs. Prices for individual parts as well as complete systems have come down considerably so that solar hot water (assisted) systems have now become economical alternatives to traditional heating systems. They have always been an ecological alternative because every cubic foot of water heated by solar energy corresponds to several kilos of carbon dioxide not released into the atmosphere.

The system shown in Figure 1 is an example of a natural – or passive – circulation system. It consists of the collector located on the roof, the solar storage tank located above the collector and the back-up tank, which feeds the shower. The system works as follows. Cold water from the mains enters the solar storage tank at the bottom, leaving it again to travel through the feed pipe to the collector. As sunlight heats the collector, heat is transferred to the water, which rises to the top of the collector from where it returns to the storage tank through the return pipe. Upon entering the storage tank, it displaces colder water in the lower part of the tank, which then flows to the bottom of the collector thus completing the loop. A pump is not necessary as the system relies upon the well-known physical properties of water: as water warms up it becomes relatively lighter and thus rises. Such a system is very reliable, relatively cheap, and easy to install. There are, however, some disadvantages: it responds rather slowly to changes in temperature and, more problematic, it is difficult to protect from freezing.

The pumped system shown in Figure 2 has better responding characteristics and can be protected from freezing. The system works as follows. Cold water from the bottom of the storage tank is pumped to the collector.

Figure 1: Passive Thermosyphon System

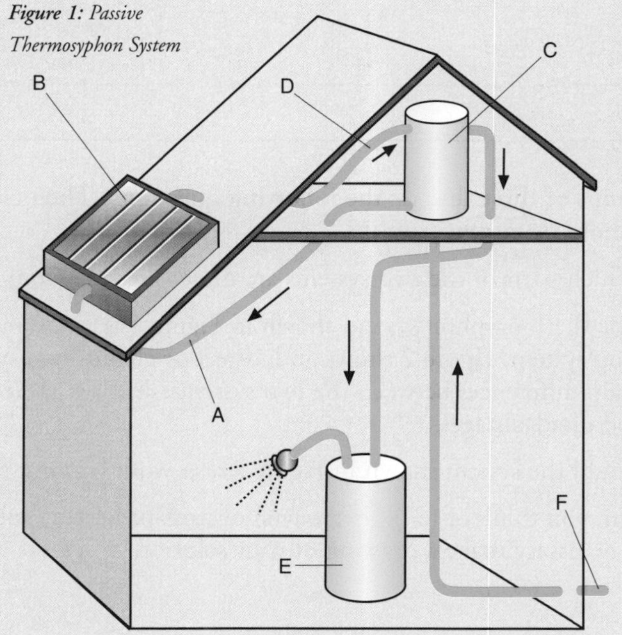

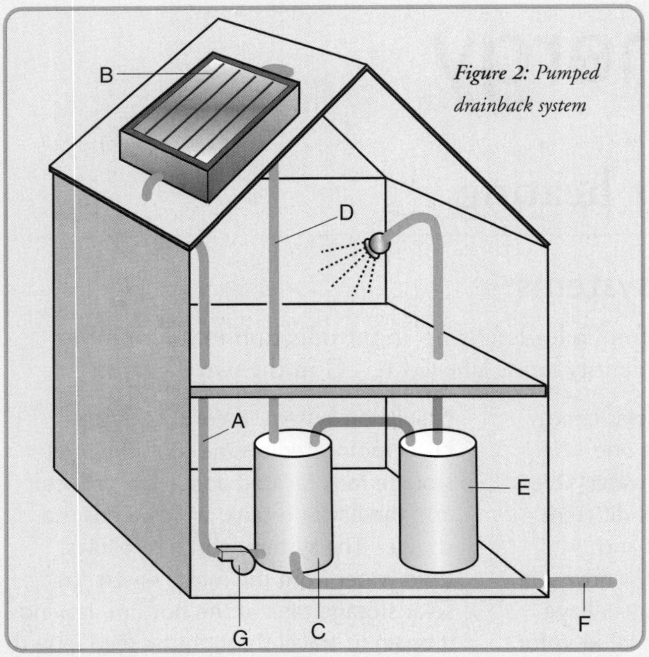

Figure 2: Pumped drainback system

Passing through the collector coils, it picks up heat and becomes gradually warmer so that it returns to the solar storage tank some ten to fifteen degrees warmer. Inside the relatively large and well-insulated solar storage tank it transfers its heat to the cold water coming from the mains. The solar hot water tank is thus not only a storage tank but also a heat exchanger which causes heat from the collector loop to be transferred to the hot water loop.

The addition of temperature control devices allows water to be circulated through the collector loop only when the temperature difference between water in the collector and water in the storage tank exceeds a certain level. Also, when the collector temperature drops to near freezing, a valve in the pump housing opens so that all the fluid in the collector loop is drained back into the storage tank. There is therefore no risk of freeze-ups.

2 VOCABULARY

Find the verbs in the text that relate either to the movement of fluid, the changes of temperature or to a more general description of the systems. Copy and complete the table below.

Movement of the liquid	Temperature	General description of the system
to circulate	to transfer	to employ
to enter	to heat	to develop
...	...	...

3 TASK

In groups of three discuss the following questions. Then compare your results with those from other groups.

1 Which parts of the two systems are under pressure from the mains?

2 The thermosyphon system shown in Figure 1 is an example of a direct or open-loop system. Figure 2 shows an indirect or closed-loop system. What are the main differences between the two systems? Explain their respective advantages and disadvantages.

3 Could the system shown in Figure 2 also work without pump G?

4 Can you think of (an)other way(s) of frost-protecting solar hot-water systems? If necessary make a drawing of your solution.

4 The pumped draindown system

READING

The following text describes the pumped draindown system, a system designed for use in areas in which sub-zero temperatures are rare. In pairs complete the piping arrangement in Figure 3 according to the information given (if possible, use overhead transparencies) and prepare an explanation of your solution. Note the words used to describe the structure of the system and the prepositions that go with them.

"... Let's now turn to Figure 3. Despite the seemingly complex arrangement of pipes the pumped draindown system is an example of a direct system, i.e. it consists of one open loop only. Water from the mains enters the system at the bottom of the solar storage tank from where the colder water is pumped to the collector. The collector is mounted on the slanted roof of the house and comprises two horizontal pipes (or headers) linked by several vertical pipes (or risers) in such a manner that the water rises slowly from the bottom to the top of the collector. Hot water then returns to the solar storage tank through the return pipe. The collector feed pipe and collector return pipe are connected by a Y-shaped piping arrangement which allows water to be drained through the joints at point A, B and C. For that purpose the system incorporates two valves fitted into the joints A and B and a frost sensor attached to the outside of the house which controls the valves.

Occasionally one also finds the drain pipe equipped with a valve. The particular system shown in Figure 3 features a back-up tank which increases the amount of hot water that can be stored and which is fitted with auxiliary resistance heater elements that provide hot water when the system is not gaining enough solar heat."

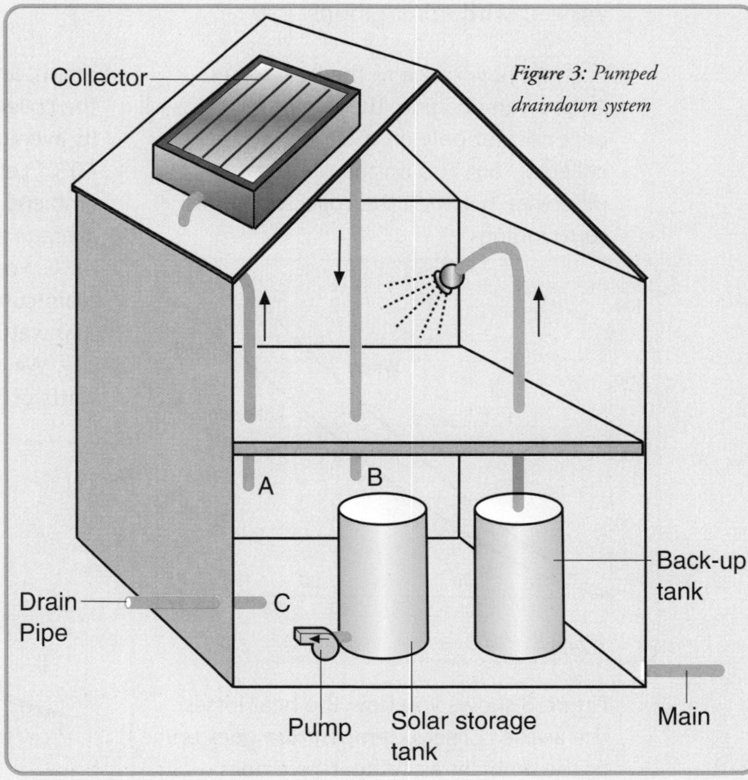

Figure 3: Pumped draindown system

Focus Solar intensity

5 The thermal efficiency of solar collectors

LISTENING

You will now hear part of a presentation. The speaker is explaining the different influences affecting the thermal efficiency of solar collectors. Before you listen, study the graph and copy the table next to it. As you listen, label the two axes of the graph and complete the table.

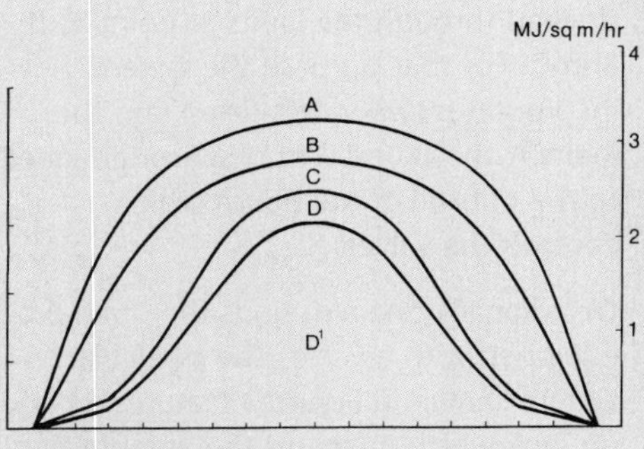

Figure A: Solar intensity

Table A: Solar intensity during one day

	Type/Weather	Maximum (W/m²)	Maximum total gain (KWh/m²)
A	tracking/ideal	...	...
B	...	...	XXX
C	...	650	...
D	...	...	...
D1	...	...	...

6 Heat loss

READING

In groups of three or four read the transcript of the second part of the presentation on solar collectors and answer the questions that follow. Then compare your answers with other groups.

"Let me now come to the heat losses. As you can imagine, the rate of heat loss depends not only on the type of collector, but also on the temperature difference between the collector and its surroundings.

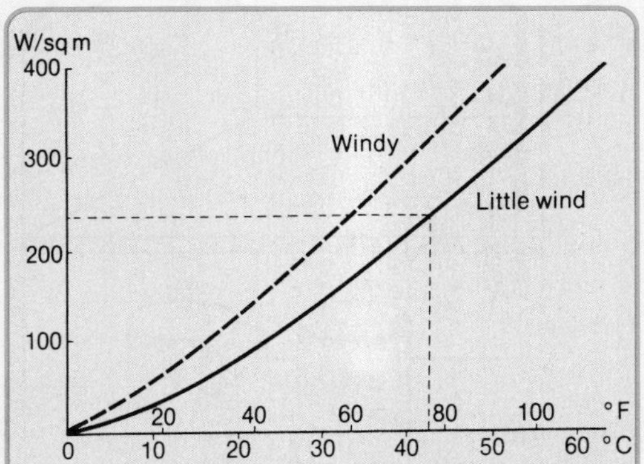

Figure B: Heat losses from a collector

Figure B shows you how the heat losses rise as the collector temperature goes up or the ambient air temperature goes down. As an example of the use of this graph, let us assume that the fluid enters the collector at 55ºC and leaves at 65ºC. Its average temperature will then be 60ºC. Let us also assume that the ambient temperature is 18ºC so that the average temperature difference will be 42ºC. For that figure we read off the main curve a heat loss of approximately 250 watts per square metre on a calm day. We can now superimpose the 250 watts of heat loss

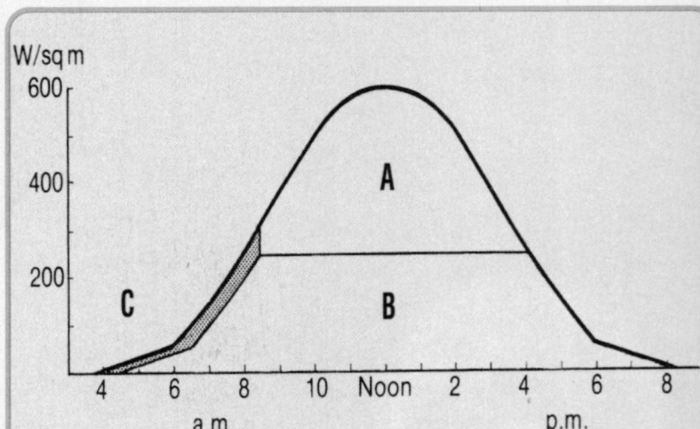

Figure C: One day's useful heat

on our curve D, and we get the graph shown in Figure C. As you can see now, a large proportion of the heat input does not become useful heat. It is wasted in heat losses. Figure C is, of course, approximate, because heat losses actually change during the course of the day. We keep them constant to avoid complication. Now, what does Figure C tell us? It tells us firstly that we start getting useful heat only after around 8 o'clock, secondly that the maximum heat gain is about 370 watts per square metre around midday, and thirdly that the system does not deliver useful heat after around four o'clock. The total heat gain is just over 2 kilowatt hours for one square meter on an average summer day. The average 'collection efficiency' is only about 40%.

As the average family in Northern Europe needs the equivalent of 12 kilowatt hours of energy per day to heat its water, six square meters of collector surface would be necessary to provide enough hot water. And that's on a summer day!

Clearly, such results are not acceptable. How can we change the system to get better results? Another look at Figure C reveals that there would be more useful heat if either the total input of heat was greater, or the losses line was lower down. Now, we can't really increase the input side, but we can change the losses side. All we need to do is to accept a somewhat cooler collector loop temperature. If, for example, we are willing to accept an average fluid temperature of 45 degrees Celsius instead of the 60º, the heat losses would only be 100 watts per square metre, which would increase collector efficiency to something like 65%. The more we go down with collector loop temperature, the more efficient the system becomes. To make it even more efficient, we can couple it with a heat pump, but I think we should have a short break before I talk about such a combination."

1. In the example given, the heat loss is approximately 250W/sqm on a calm day. How much higher is the loss on a windy day?
2. Label the three areas A,B,C in Figure C with "Heat losses", "Warming up" and "Useful heat" and prepare an explanation in your own words.
3. How does the speaker explain the higher efficiencies of "low" temperature collectors?
4. The speaker says that efficiency also depends on the type of collector. What different types do you know? Can you make a rough drawing and explain how they are built?
5. The speaker says that Figure C is approximate. Why is it approximate? Can you draw a more realistic graph and prepare an explanation for the change?

PHRASES

Expressing contrast

Although the sky was clear, it was very cold.
The total heat gain is 12 kilowatts an hour, **but** much less in winter.
Unlike a tracking system, a flat fixed collector cannot be moved.
Whereas/While a tracking system receives more heat, it costs more initially.
A tracking system and a clear sky produce impressive results, clear skies, **however**, are rare.
Despite optimum conditions, collection efficiency is not very high.
The additional cost is considerable, the heat gain **in contrast** minimal.

7 PRACTICE

Which of the two words in brackets is best?

One-loop solar hot-water systems are very cheap and reliable. ...[1] *(Unlike/Although)* two-loop systems, they do not require a water pump because they rely solely on the physical properties of water. But ...[2] *(although/despite)* their simplicity, their use is generally restricted to heating water for swimming pools because they are difficult to protect from freezing. ...[3] *(Whereas/In contrast)* to a one-loop system, a two-loop system can be protected from freezing. ...[4]*(However/While)*, two-loop systems without anti-freeze must be equipped with frost sensors and valves. ...[5]*(Whereas/But)* two-loop systems without anti-freeze liquids present no hazard to the environment, non-freezing systems must be carefully protected against leaks because many anti-freeze liquids are toxic. ...[6] *(Although/Despite)* modern solar hot water systems are very reliable, they are not used very frequently in Western Europe. Many people consider the installation costs for solar heating to be too high. ...[7]*(While/Whereas)* it is true that solar hot-water systems increase the price for a hot-water system, they have proved to be cost effective in the long run – and that not only in southern countries.

8 WRITING

Figure D shows a tidal power station as proposed by the Spanish inventor Antonio Ibanez de Alba. Use the information given in the drawing and in the sentences below for a description of the process. Use expressions from section 4 (e.g. *consist of*, *incorporate*, *be equipped with*, etc) and linking words like *as*, *then*, *while*, *during*, *as long as*, etc.).

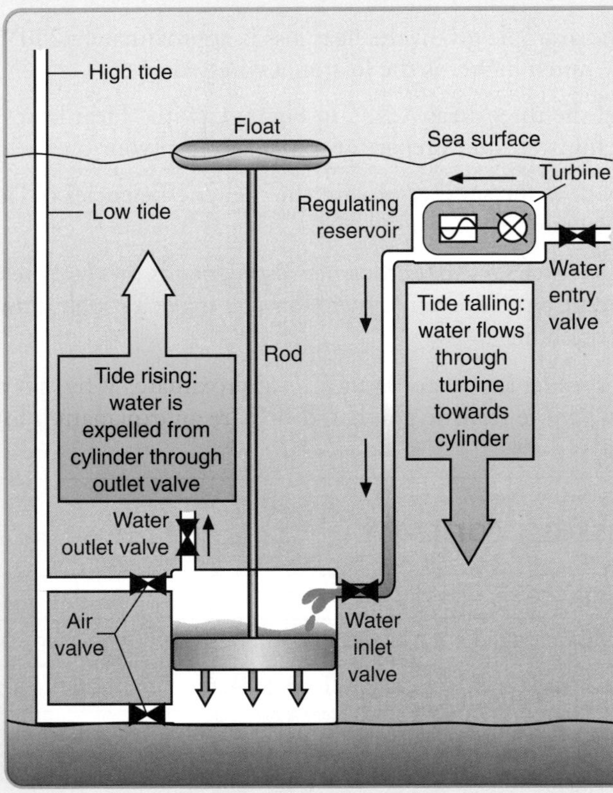

The tide falls.
- The float/piston moves down.
- Water inlet valve and lower air valve are opened.
- Air is forced out through the snorkel.
- Water passes through the turbine.
- Electricity is generated.

The tide rises.
- The float/piston rises.
- Water inlet valve is closed.
- Water outlet valve is opened.
- Lower air valve is opened.
- Air at atmospheric pressure enters the chamber.

GRAMMAR

Participle clauses

1 a When Mr Singh came back from a meeting, he found a letter from his secretary.

b **Coming** back from a meeting Mr Singh found a letter from his secretary.

2 a After the hot water leaves the storage tank, it is mixed with cold water.

b After **leaving** the storage tank the hot water is mixed with cold water.

In sentences 1a and 2a the subjects of both parts of the sentences are the same. The sentences can be shortened using a participle.

9

PRACTICE

Which of the following sentences can be shortened as explained above? How?

1 As the water rises through the collector, its temperature increases.
2 After the water has passed through the collector, it returns to the solar storage tank.
3 As soon as the pump is activated, water starts to circulate through the collector loop.
4 When the warm water returns to the storage tank, it displaces colder water at the bottom.
5 When the differential thermostat detects a positive temperature difference of >10 degrees, the pump is activated.
6 When hot water is drawn from the top of the storage tank, cold water automatically enters the storage tank.
7 When the water passes through the turbine, it imparts its energy on the turbine blades.
8 When the water enters the chamber above the piston, it forces the air out of the air valve.

MODULE 2b

Atomic power

Focus Authentic reading

1

READING

A pressurized-water reactor

Read the text describing the water cycle in a nuclear power plant, and complete the diagram below: what are the parts labelled A – D and the figures for temperatures labelled **a – f**?

The principle of the pressurized-water nuclear power plant is illustrated below. Fuel rods for this reactor consist of uranium dioxide, enriched to about 3 per cent in uranium-235, hermetically sealed in tubes of a zirconium alloy. This zirconium tubing constitutes the first barrier against escape of highly radioactive fission products that form in the fuel as the end product of the nuclear reaction. Assemblies of these zirconium-clad fuel rods are mounted in a heavy-walled steel pressure vessel and are surrounded by flowing water entering at a temperature of around 540ºF and leaving around 600ºF, held at a pressure of around 2250 pounds per square inch to prevent boiling. The water slows down neutrons produced in fission and increases their probability of reacting with uranium-235 to such an extent that the uranium fuel constitutes a critical mass capable of sustaining a nuclear chain reaction. To hold the chain reaction at a steady rate, a variable amount of neutron absorbing boron is used in the reactor, partly as movable control rods and partly as boric acid dissolved in the water. Pressurized water is pumped through the reactor by the circulating pump, past the gas-cushioned pressurizer, which holds the pressure constant, and through the steam generator. There heat is transferred from the primary pressurized water at

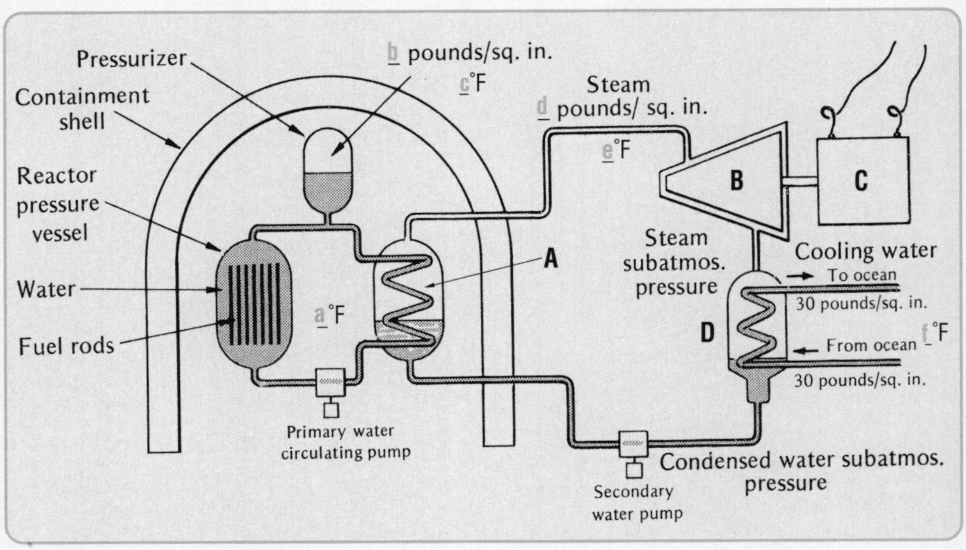

600ºF and 2250 pounds per square inch to secondary water boiling at a lower pressure of around 720 pounds per square inch to make steam at around 506ºF.

The steam flows through a turbine driving an electric generator and then passes to the condenser, where it is condensed at subatmospheric pressure. The condensate is returned to the steam generator by the condensate pump.

In the condenser, heat from the condensing steam is transferred through cooling coils to cooling water at a pressure above atmospheric, which leaves the condenser at a temperature typically 20ºF warmer than incoming water. In some plants, this cooling water is drawn from the ocean or other natural sources; in others, it is recirculated through cooling towers. Disposal of the heat contained in this warm water without adverse effect on the environment is one of the problems of all steam-electric plants, nuclear as well as fossil. In light-water-nuclear plants, however, about 50 per cent more warm water must be handled than in an efficient fossil-fuelled plant, because the thermal efficiency of the water-cooled nuclear power plant is only 32.5 per cent due to its relatively low steam temperature of 506ºF, compared to a fossil-fuelled plant with a thermal efficiency of over 40 per cent obtainable from steam temperature around 1000ºF. The environmental impact of warm water from a nuclear power plant can be dealt with satisfactorily either by siting on a natural body of water with adequate heat absorption capacity, such as the ocean or a large river, or by using cooling towers. The cost of heat disposal from a water-cooled nuclear plant is, however, somewhat higher than from a fossil-fuelled plant.

GRAMMAR

Reduced relative clauses II

Disposal of the heat (which is) **contained** in this warm water is only one of the problems of nuclear power stations.
The heat is transferred to secondary water (which is) **boiling** at a lower temperature.

- Note how the relative pronoun and the form of *be* is omitted in these sentences, leaving just a participle. *(See also Module 1, p. 83.)*
- Because they are shorter, such sentences are often found in technical writing.

2 PRACTICE

Make true sentences from the sentence switchboard.

1 The fuel rods	illustrated in the diagram	is of the pressurized-water type.
2 The neutrons	mounted in the pressure vessel	are surrounded by water.
3 The steam	condensed at subatmospheric pressure	has gained approximately 20ºF.
4 The heat	leaving the power station	are slowed down by the water.
5 The nuclear power plant	added to the cooling water	maintains a steady chain reaction.
6 The boric acid	produced by nuclear fission	is recycled by the condensate pump.
7 The water	dissolved in the water	must be dissipated.

3 DISCUSSION

Discuss the pros and cons of nuclear power.

For the last twenty years there has been an ongoing public debate about the benefits and dangers of nuclear power. The threat of global warming as a result of burning fossil fuels (to produce energy) has made that discussion even more complex.

Split up into two groups and collect arguments For *(Group A)* and Against *(Group B)* nuclear power. Take approximately 15 minutes to collect your ideas. Select a chairperson to lead the discusssion and start the discussion.

4 VOCABULARY

Which of the words given is best for each gap?

consist of/comprise • contain • constitute

1. The primary water cycle … four main parts: the reactor, the pressurizer, the condenser and the water pump.
2. The containment shell … a final barrier against the escape of radioactivity.
3. The primary water … a certain amount of dissolved boric acid.

to raise • to rise

4. If water in a tank is heated, the hot water will … to the top while the relatively colder water will sink to the bottom of the tank.
5. In the reactor the water temperature is … to about 540ºF.

mass • amount

6. The … of boric acid that must be dissolved in the water will depend on the kind of fuel rods used in the reactor.
7. For a chain reaction to take place a critical … of uranium fuel is required.

prevent • protect

8. To … any contamination, water from the primary water cycle must not come into contact with secondary water.
9. The containment shell serves to … the reactor from outside as well as the environment from the escape of radioactivity.

Focus Learning vocabulary

5 Recording vocabulary

STUDY SKILL

How do you record new English words? Which of the following suggestions do you like best? What are the advantages and disadvantages of each method?

1 You can record new words as a list in a vocabulary notebook (preferably one small enough to take with you on a bus, train etc.).

steam = Dampf
pressure = Druck
fuel = Brennstoff

2 With some words it may be helpful to record them as word families:

water, steam, vapour, liquid
nuclear energy, fossil fuel, solar energy, water power

You can do this as a table, like this:

Verb	**assemble**	**react**	**consider**
Noun	assembly	reaction	consideration
Adjective	—	—	considerable
Agent	assembler	reactor	—

3 You may use index cards. The advantage of index cards is that you can write the word or the expression on one side and the translation on the other side.

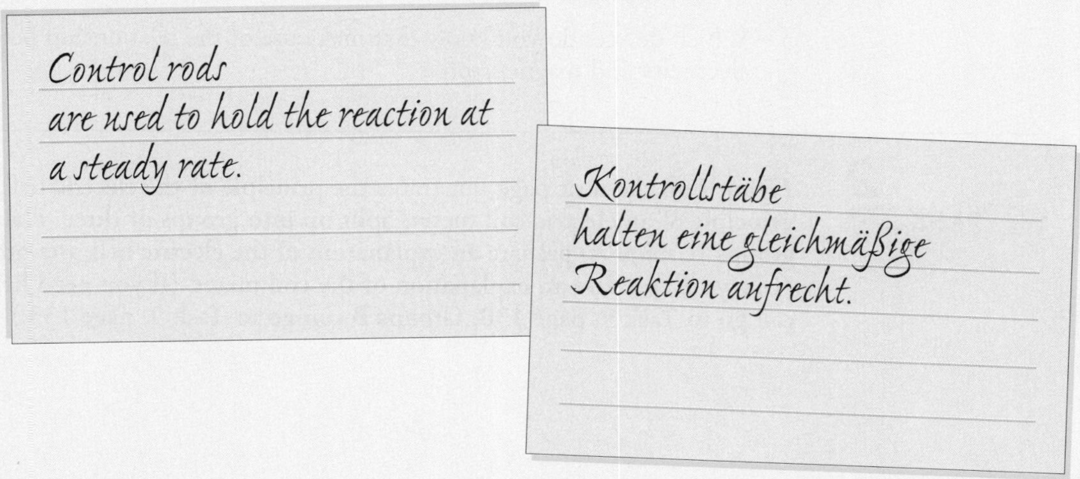

Keep the cards in a box with three partitions. Store all the new expressions in the first partition. If you still know the expression the next time you study vocabulary, place the card in the second partition. If you have forgotten it the next time, the card goes back into the first partition. If, however, you still know the expression, the card goes into the third partition. In addition to learning and recording new words, check the second partition regularly. Check the third partition at longer intervals of about a month.

MODULE 3a

An electric motor

Focus Magnetism and electricity

1 READING

An experiment

Magnetism and electricity are natural phenomena that have interested mankind for many centuries. The text below describes an experiment which was carried out in the middle of the 19th century to demonstrate the relationship between magnetism and electricity. Read the text and together with your partner make a drawing of the experiment. Discuss the group results in class.

A rectangular frame of stiff wire hangs by means of two flexible wires so that the bottom of the frame is parallel and close to a fixed horizontal wire. Frame and fixed horizontal wire are connected so that when a button is pushed, a current of approximately 20 A flows through them in opposite directions. The battery needed for the experiment is a low-voltage, single-cell battery. The resulting magnetic forces cause the frame to swing away from the fixed wire. The movement is small but quite distinct. A sheet of transparent paper fixed across the frame acts as a damping element and controls the movement.

- **Now discuss the following questions.**

1 What is the reason for the movement?

2 What happens if the polarity on the battery is reversed?

3 Which devices do you know that make use of the relationship between electricity and magnetism?

2 TASK

***Figure 1* on the next page illustrates the principle of the electric bell, *Figure 2* the principle of an electric coil meter. Split up into groups of three. Half of the groups *(Groups A)* prepare an explanation of the electric bell, the other half *(Groups B)* prepare an explanation of the coil meter. (If you need help, Groups A can go to Task 9, page 130, Groups B can go to Task 9, page 133.)**

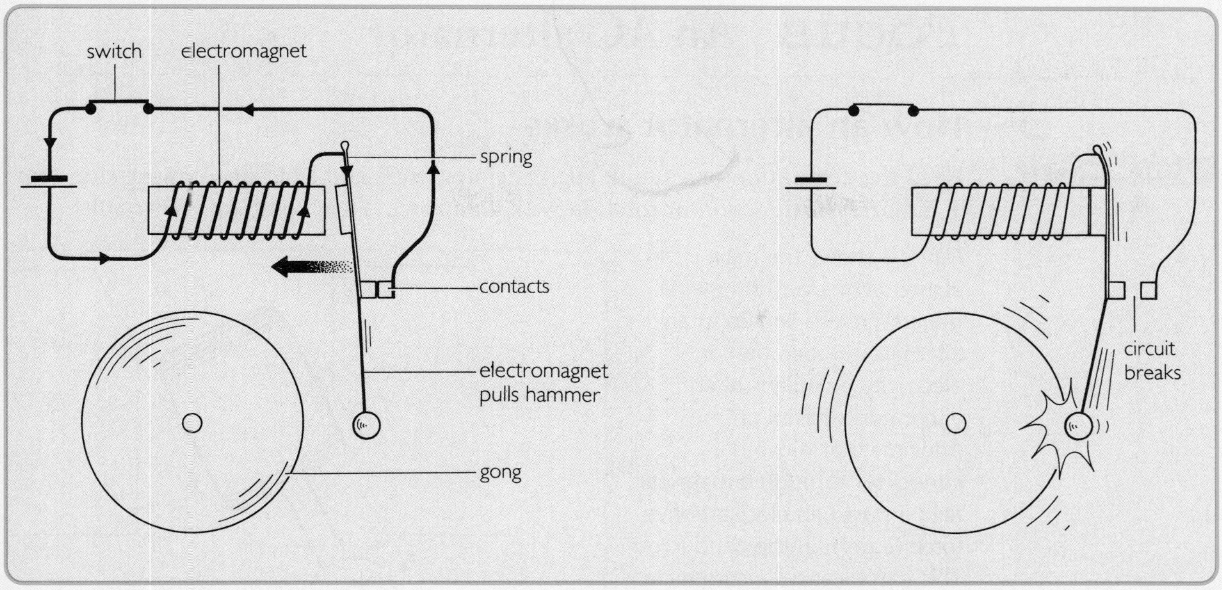

Figure 1: The electric bell

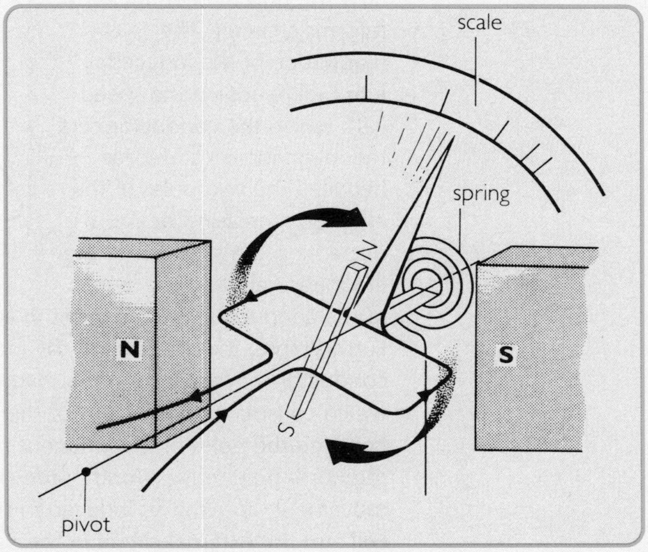

Figure 2: The electric coil meter

- Now work in pairs, with one participant from group A and one from group B. Explain the purpose and function of the electric bell/the coil meter to one another. Also explain how the working principles relate to the experiment described on the previous page.

Focus An AC alternator

3 How an alternator works

READING

Read the description of a simple AC generator, copy and label the drawing (for item 1 indicate the direction of rotation with an arrow). Then compare your results.

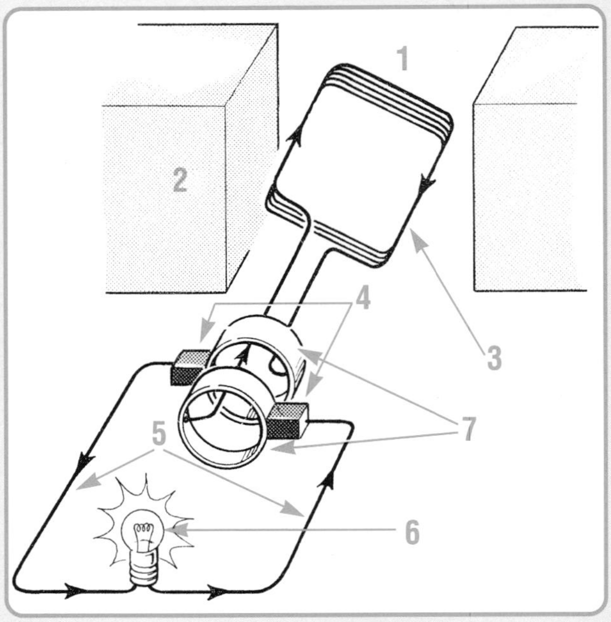

Figure 3: A simple AC alternator

Figure 3 shows the main elements of a very simple AC generator, also known as an alternator. Generation of electricity by means of an alternator is based on the principle that moving a conductor through a magnetic field induces an electromotive force (e.m.f). in the conductor. This e.m.f. causes a current to flow through the conductor if it forms a circuit. The magnitude of the induced e.m.f. depends on the speed with which the conductor cuts the magnetic lines of force between the two poles of the magnet. Increasing or decreasing the strength of the magnetic field and the length of the conductor will also result in an increased or decreased current in the wire. Furthermore, the direction of the induced current depends on the direction in which the conductor moves through the magnetic field. In *Figure 3* the alternator contains a coil, which consists of many turns of thin wire. The coil is rotated in a clockwise direction between the poles of a permanent magnet. As shown in the diagram, the side close to the north pole moves up, the side close to the south pole moves down. As a result of this movement, an e.m.f. is induced in the coil which causes a current to flow through the coil and the external circuit in the direction shown by the arrows, causing the light bulb to light up. Coil and outside circuit are connected by two carbon brushes which rub against two copper slip rings fixed to the two ends of the coil. The current induced in the coil is not stable but varies between a positive and a negative maximum. It is strongest when the coil is in a horizontal position relative to the magnetic lines of force. This is because at that moment the wires cut the magnetic field at right angles and therefore at maximum speed. One quarter of a turn later the induced e.m.f. and current are zero, because for this brief moment the two sides of the coil are now travelling parallel to the magnetic field and no field lines are being cut. Another 90 degrees later, maximum current is induced again, but now the direction of current flow is reversed, because the side of the coil which was moving upwards is now moving downwards through the magnetic field.

4 TASK

Discuss the following statements. Correct the ones that are wrong, and explain why they are wrong.

1 Outside circuit and coil circuit are connected in parallel.

2 Increasing the speed of rotation of the coil increases the e.m.f. and the strength of current produced as well as the frequency of the current generated.

3 The carbon brushes are soldered to the two ends of the coil.

4 The two slip rings must be insulated from one another.

5 The more windings the coil carries, the more force is required to rotate the coil in the magnetic field.

6 No e.m.f. is induced if the coil rotates counter-clockwise.

7 The light bulb will burn intermittently.

5 VOCABULARY

What is the difference between the following verbs? Divide into six groups. Each group takes one group of verbs and makes one sentence with each verb.

1	to contain	to consist of	to include
2	to turn	to rotate	to spin
3	to cause (to do sth.)	to depend on	to result in
4	to connect	to fix	to mount
5	to change	to reverse	to return
6	to induce	to exert	to increase

6 LISTENING

Coil position and current

Figure 4 shows how the current generated in the coil varies during a complete rotation of the coil. In his explanation of *Figure 4*, the speaker makes four mistakes.

Listen to the explanation, make a list of the mistakes and correct them.

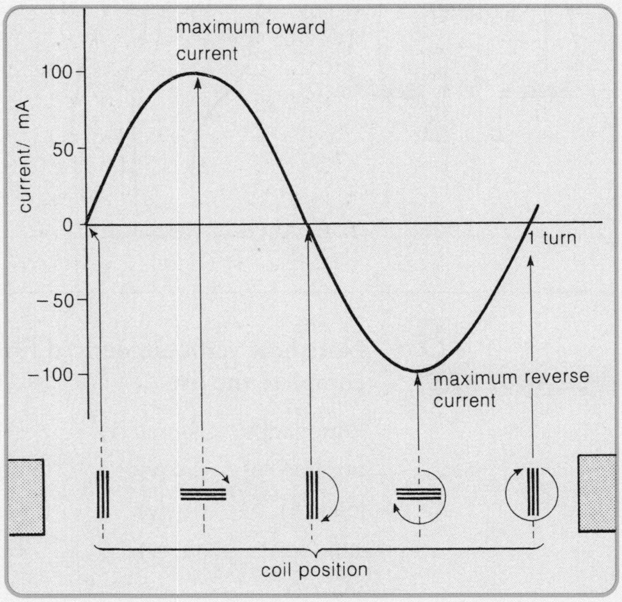

Figure 4: Current and coil position

7 TASK

Decide which drawings belong to which descriptions, and then put them in the right order. Work with your neighbour.

A

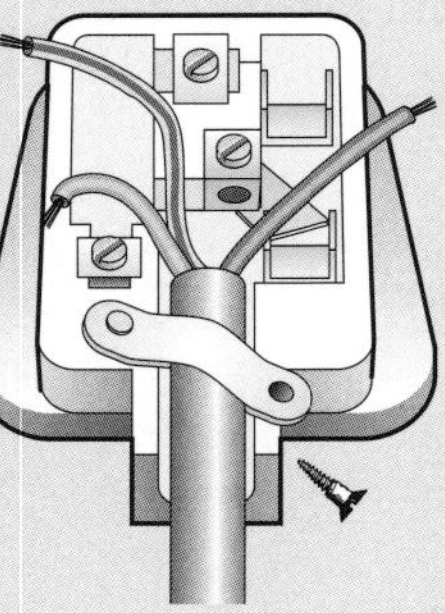

B

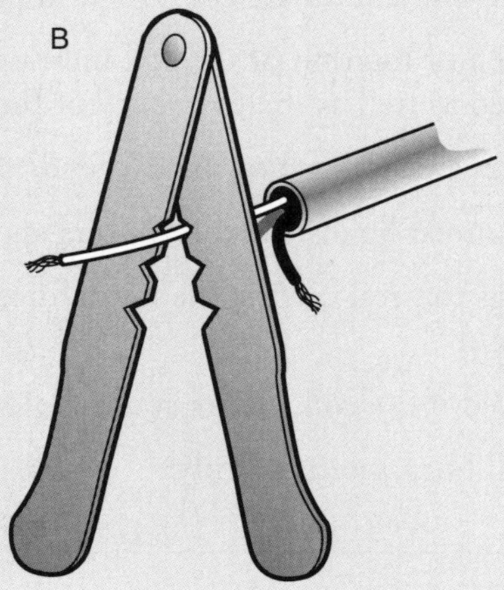

C

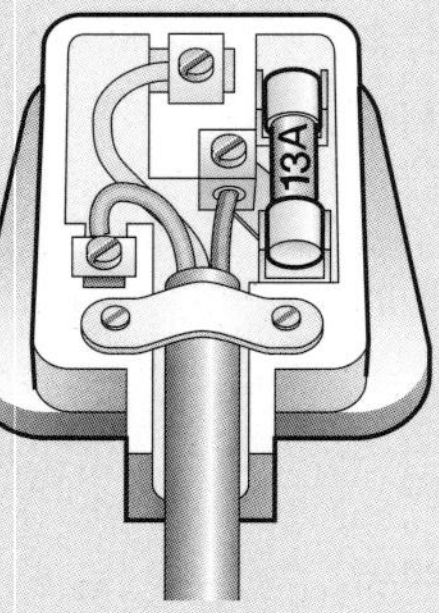

F

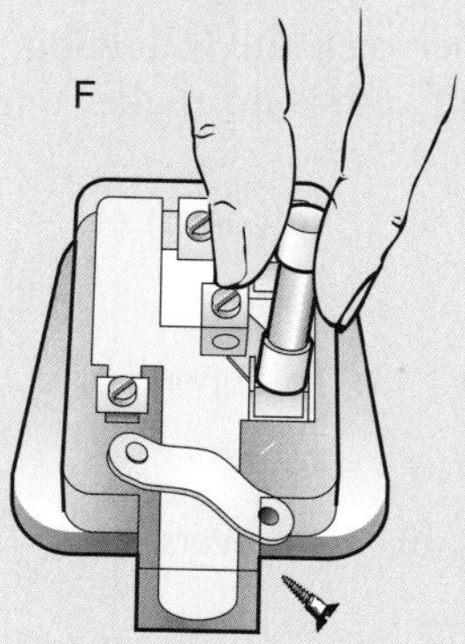

D

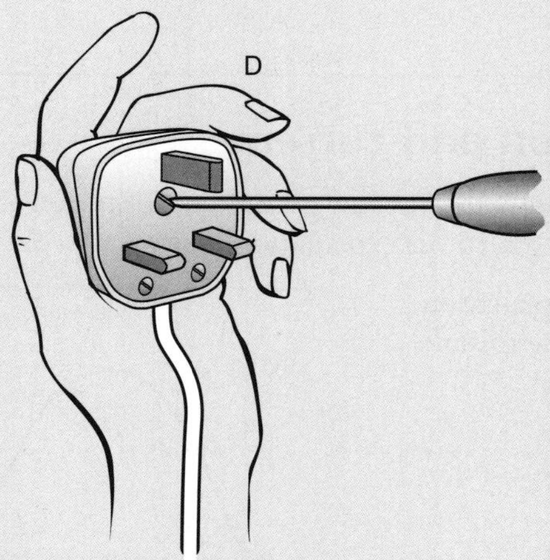

E

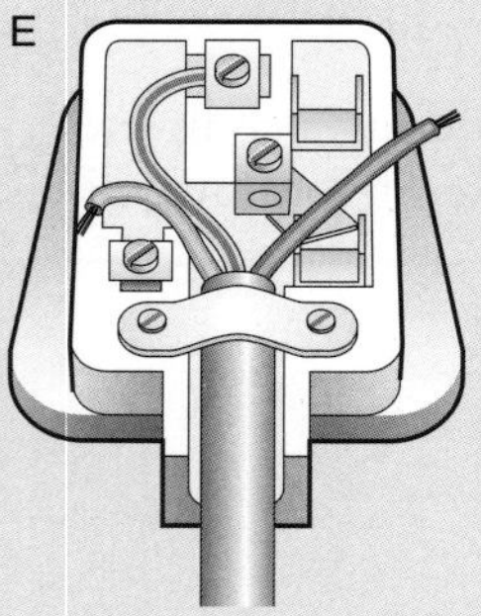

1
Unscrew the plug cover. Loosen one flex clamp/cord grip screw, remove the other.

2
Finally check:
1 wires are connected to correct terminals;
2 there are no stray 'whiskers' of wire;
3 flex clamp is on outer sheath of cable, not on wires;
4 all screws are tight.
Then:
5 fit the correct fuse 6 refit plug cover and screw tight.

3
Remove the fuse. Carefully lever it out with a screwdriver if necessary. Loosen terminal screws.

4
Position flex in plug, cut the wires to reach about 13mm (1/2 inch) beyond each terminal. It may be necessary to carefully cut away flex outer sheath.

5
Fasten the cable clamp firmly over the outer sheath. Fit each wire into the appropriate terminal hole and tighten each screw. Check that there are no stray 'whiskers' of bare wire.
NB: If the wires are small it may help to fold them over and put them into the terminal holes.

6
Carefully strip enough insulation to expose about 6 mm (1/4 inch) of wire for screwhole terminals, about 13 mm (1/2 inch) for clamp type ones. Take care not to cut any strands of wire. Twist the strands of each wire together.

8 VOCABULARY

Note how verbs are derived from adjectives and nouns. Use your dictionary to complete the list.

loose (adj)	*loosen* (v)	short (adj)	... (v)	strength (n)	... (v)
position (n)	... (v)	tight (adj)	... (v)	lever (n)	... (v)
load (n)	... (v)	weak (adj)	... (v)	screw (n)	... (v)
stiff (adj)	... (v)	clamp (n)	... (v)	grease (n)	... (v)
seal (n)	... (v)				

MODULE 3b

Instructions

Focus Authentic reading

1

Replacing fuses

READING

Read the description of how to replace a fuse. Would you have to do the same in your country? Are your appliances protected in a similar way? What do you do to check a fuse?

Replacing fuses

Plug Fuses
If an appliance stops working, switch off immediately (turn off appliance itself and socket outlet if both have switches). Remove plug, unscrew top and take out fuse.
Fit a substitute fuse or check existing fuse as described below and replace if necessary. Also check connections for loose or disconnected wires. Refit plug cover. If the appliance still fails to work, and the main fuse is intact, get the appliance checked by an approved electrician.

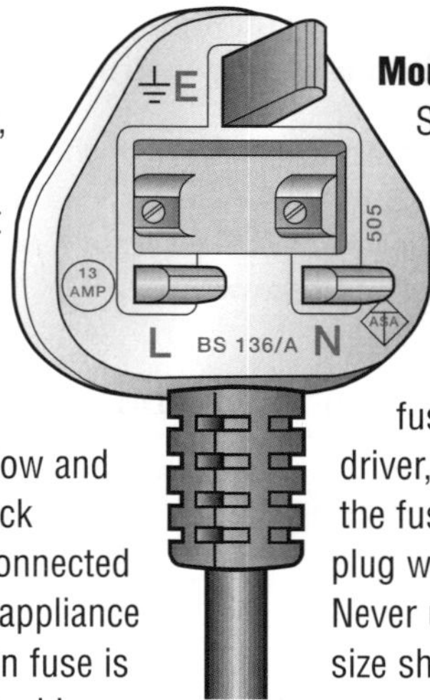

Moulded plugs
Some appliances have a ready fitted plug of the non rewirable type.
To replace the fuse, switch off and unplug the appliance. Prise the fuseholder out with a screw-driver, replace the fuse and refit the fuseholder. Do not use a plug without a fuseholder.
Never use a fuse greater than the size shown on the plug.

Faulty plugs
Should a moulded plug become damaged or faulty, you should cut it off and make sure it is disposed of properly. Do not leave it where children may find it and plug it into a supply socket.

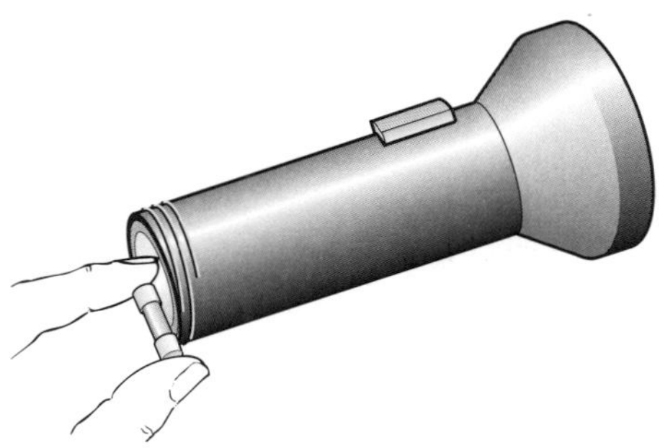

Checking fuses
Remove base of a metal cased torch. Place one end of fuse on bottom of battery and other end on torch casing. If bulb fails to light when torch is switched on, the fuse has blown.

GRAMMAR

Omission of definite article

Place one end of **fuse** on **bottom** of **battery**.
Place one end of **the fuse** on **the bottom** of **the battery**.

- To shorten a text, technical writers often leave out the definite article, especially in instructions.

2
PRACTICE

Read the text in section 1 again and decide where it would be grammatically correct to add a definite article.

3
VOCABULARY

Repair work usually involves three steps: Inspection – Disassembly – Assembly. The following verbs are often found in repair manuals. Which would you expect to find in instructions for inspection, which in disassembly instructions and which in assembly instructions. (Some verbs may fit more than one category.)

adjust • align • analyse • check • connect • disconnect • evaluate • fasten • grease • inspect • install • loosen • lubricate • make sure • measure • mount • open • place • position • pull off • release • remove • replace • slide down • support • test • tighten • unscrew

Inspection	Disassembly	Assembly
analyse	disconnect	adjust
…	…	…

FOCUS Tests

4 Preparing for international exams

STUDY SKILL

International exams often use multiple choice items to check vocabulary or reading comprehension.

Read the following text about a DC generator. Decide which of the words or phrases on page 96 best fit the numbered gaps.

The construction of a simple DC generator shown in Figure 1 is as follows. The single armature coil is located in slots in a cylindrical piece of soft steel called the armature coil (not shown). The coil shown is a single loop of wire, but in reality it ... (1) many loops of thin wire. The ends of the coil are soldered to semicircular segments to form the commutator. The two segments are ... (2) each other. Electrical contact between the armature coil and the outside circuit is accomplished ... (3) stationary copper brushes pressed against the commutator by springs. For clarity, the springs are not shown. The magnetic field is produced not by a permanent magnet but electrically by coils of wire ... (4) the iron field poles. The two coils for the field poles are ... (5) and are connected to the brushes.

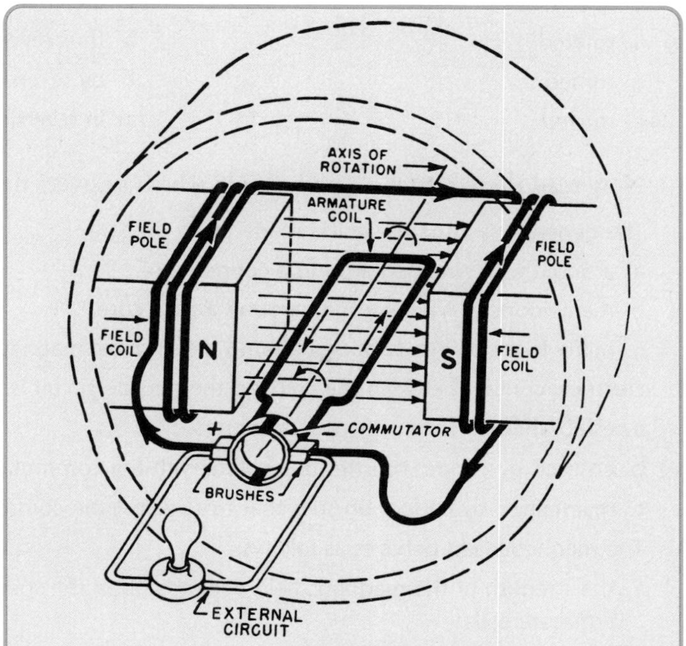

Figure 1: A simple DC generator

When the armature coil is first ... (6), a slight voltage is produced, causing a small current to flow through the field coils. This current ... (7) the strength of the magnetism between the field poles and in this way, in a few seconds, the voltage in the generator is built up to its normal value and the generator is ready for use.

With the coil in the position ... (8) Figure 6, the left-hand side of the armature coil cuts the magnetic lines of force in a ... (9) direction, and the right-hand side moves up through the magnetic field. Therefore, current flows out into the ... (10) circuit through the positive brush and returns by way of the negative brush to the armature coil. One half a revolution later, the left and the right-hand sides ... (11) and the two sides now cut the magnetic lines of force in the opposite direction ... (12) the direction of the current produced. However, the two commutator segments have also changed places so that the current still continues to flow in the same direction through the external circuit.

1 a consists of
b is constructed of
c is produced by

2 a separated from
b insulated from
c isolated from

3 a with the help of
b by means of
c through

4 a wound around
b turned around
c put around

5 a in series
b in parallel
c together

6 a rotated
b turned
c moved

7 a increases
b increasing
c increased

8 a shown in
b showing in
c show in

9 a downward
b upward
c lateral

10 a internal
b external
c outward

11 a have changed places
b had changed places
c are changing places

12 a thus reversing
b by reversing
c in reversing

● **Now read these sentences and decide which answers fit best.**

1 The generator's armature coil is made of
a a single wire wrapped around a copper core.
b many loops of wire wrapped around a steel core.
c many loops of wire wrapped around a cylindrical magnet.

2 Electrical contact between the coil and the outside circuit is
a established by means of copper springs.
b ensured by copper brushes that rotate with the commutator.
c maintained by copper brushes that rub against the commutator.

3 The magnetic field behaves as follows:
a the strength of the magnetic field depends upon the speed of rotation of the armature coil.
b the magnetic field takes a few minutes to reach its maximum.
c the electricity needed to create the magnetic field is provided by the external circuit.

4 During a full rotation of the armature coil
a the magnetic lines also rotate once.
b current fluctuates a little but flows in one direction.
c the commutator segments contact each brush twice.

MODULE 4a

Structures

Focus Basic structures

1 TASK

The drawing shows a bird's feather. What are the main properties of a feather? How would you describe its design (appearance, components)? What is the purpose of the ribs? Which side of the quill do you think points upwards in flight?

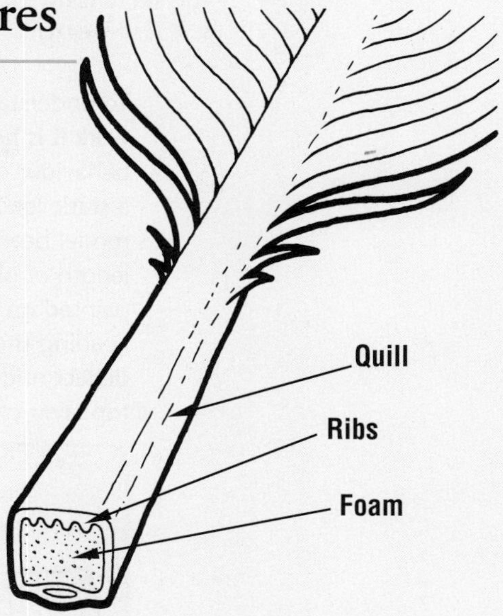

2 VOCABULARY

Look up the meaning of the following words in your dictionary, then decide which words best fill the gaps in the text. Make sure you always use the correct form.

to evolve • to develop • to design • to provide • to resist • to support • to subject to • to withstand • to rise • to span

Natural structures and those designed by humans

We are surrounded by structures that have either … (1) in nature or have been developed by human beings to … (2) loads or forces. The human skeleton, for example, is a fascinating example of a superbly "…" (3) structure. It … (4) the static load of the body mass just as well as its dynamic load. Even when … (5) such extreme dynamic loadings as are caused by an athlete the skeleton seldom breaks.

A bird's feather … (6) extremely high wind forces without breaking – despite the proverbial lightness of its wings. The reason for this is because the quill of a feather is in fact a long rectangular box with a foam core which … (7) an enormous structural strength.

Structures which have been developed by human beings have become considerably stronger and at the same time lighter. The height to which modern buildings can … (8) and the length that bridges are made to … (9) has constantly been increased as new materials have been … (10) and innovations in computing have simplified the carrying out of complicated mathematical calculations.

3 Beams

READING

Copy the sketches of the two beams. Then read the text, and complete and label the sketches with the words on the right.

Load
Area under tension
Area under compression
Adhesive tape
Neutral axis
Steel rods
Grid

To understand how basic structures work it is helpful to analyze the behaviour of a simple beam under a static load. For this purpose a model beam is made out of a length of plastic foam and a grid is painted on its side.
Loading the beam now causes it to deflect under the load putting the top layer of the beam into

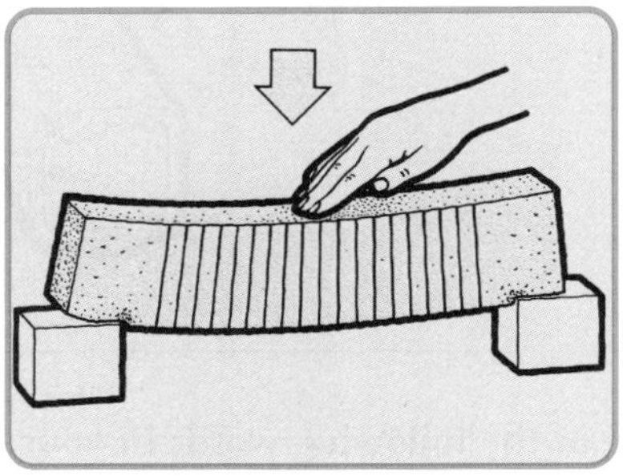

compression while the bottom of the beam is put into tension. This is clearly shown by the changes occurring in the grid. The grid spacing along the top of the beam narrows while the spacing at the bottom of the beam widens, indicating that this lower part of the beam is being stretched. The grid spacing along the centre, however, remains roughly the same. An imaginary line running along the centre of the beam is therefore described as the neutral axis: it maintains its length when the material above is compressed and the material below is stretched. The performance of the beam can be improved by preventing it from stretching at the bottom. A length of adhesive tape running the length of the beam on the bottom will do this because the tape will not stretch as easily as foam.

A similar strategy is employed for concrete beams used as structural elements in a building. Concrete is very strong in compression but weak in tension. Steel rods, on the other hand, are very strong in tension but unstable in compression.
Therefore, concrete beams are reinforced by steel rods cast in the concrete, the steel rods providing the necessary tensile strength and the concrete giving the beam the required compressive strength and stability.
Beams used in larger structures take many different forms. Some are simple solids, some are hollow, and others are given special cross-sections for strength and rigidity. In nature the quill of a bird's feather is a good example of a box-section beam. The top of the box is made stronger with ribs so

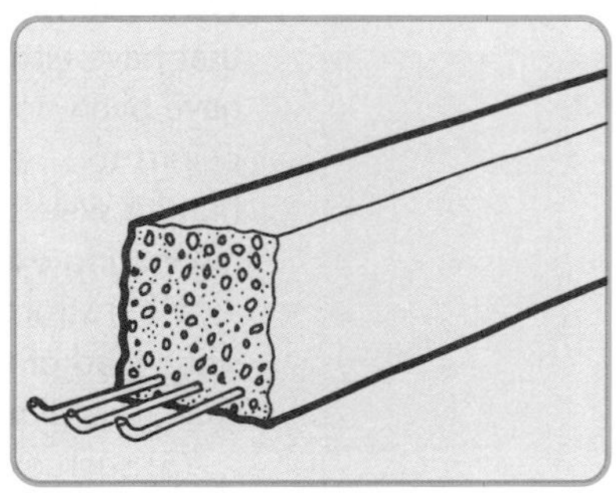

that, as the feather bends upwards in flight, it will not crumble under compression.

PHRASES

Giving definitions

… is the opposite of …/is like …/behaves like …
… consists of …/contains …
… is the general word for …/a more specific word for …
… allows …/is used to …
… is a material/is an object/is a part of …
… can be defined as …
… (maintains/remains …) when you …

4
TASK

Work in pairs. Complete the crossword below with your partner's help. Each of you has an incomplete version of it. There are no clues – you have to give your partner definitions of the words in your version (don't read the words out!), and complete the rest of the crossword from the definitions your partner gives you.

A: stay with this page. B: Go to Pairwork Task 10, page 133.

Puzzle A

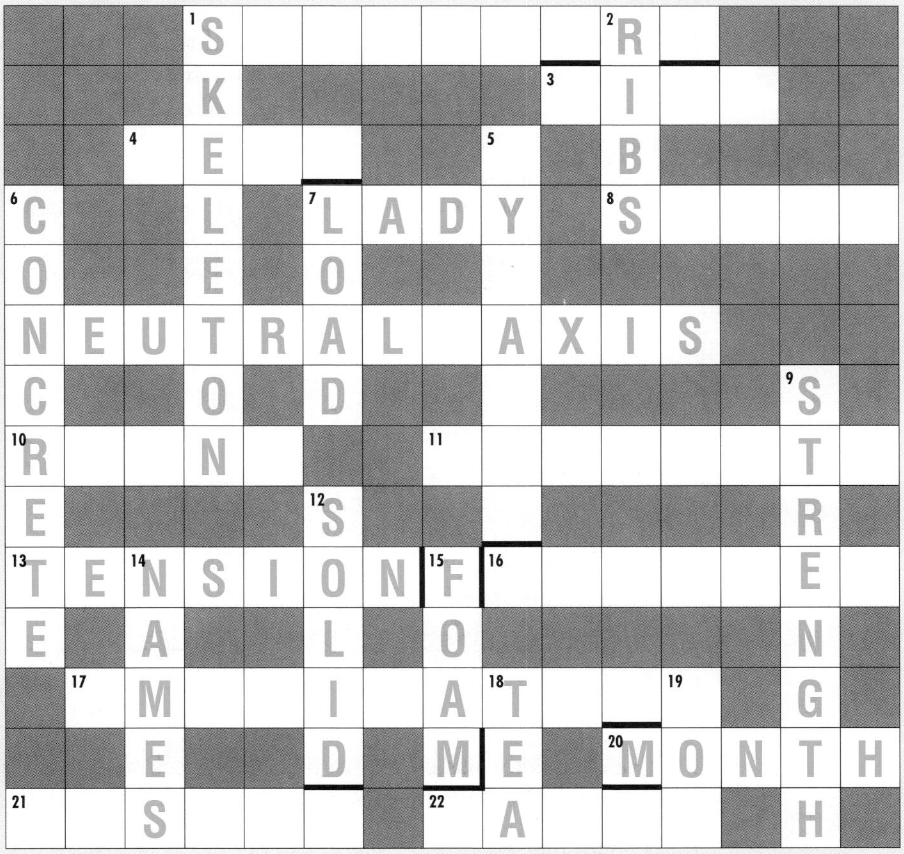

GRAMMAR

-ing clauses

1 The spacing at the bottom of the beam widens, **indicating** that this lower part of the beam is being stretched.
= The spacing at the bottom of the beam widens. This indicates that the lower part of the beam is being stretched.

2 Concrete beams are reinforced by steel rods cast in the concrete, the steel rods **providing** the necessary tensile strength.
= Concrete beams are reinforced by steel rods cast in the concrete. The steel rods provide the necessary tensile strength.

- Technical writers often combine two sentences using an *-ing* clause.
- In sentence 1 the *-ing* clause expresses the consequence of what has been said before. Both clauses have the same subject. Often the conjunctions *thus* or *therefore* are added for clarity *(... beam widens, thus indicating that ...)*.
- In sentence 2 the *-ing* clause does not express a consequence of the first sentence. It provides additional information. The subject of the second clause (here: *the steel rods*) is usually different and must be stated. Often the word *with* is added *(... in the concrete, with the steel rods providing ...)*.

5
PRACTICE

The drawing illustrates the construction of the central span of a bridge. Shorten the following sentences describing the construction of this central span.

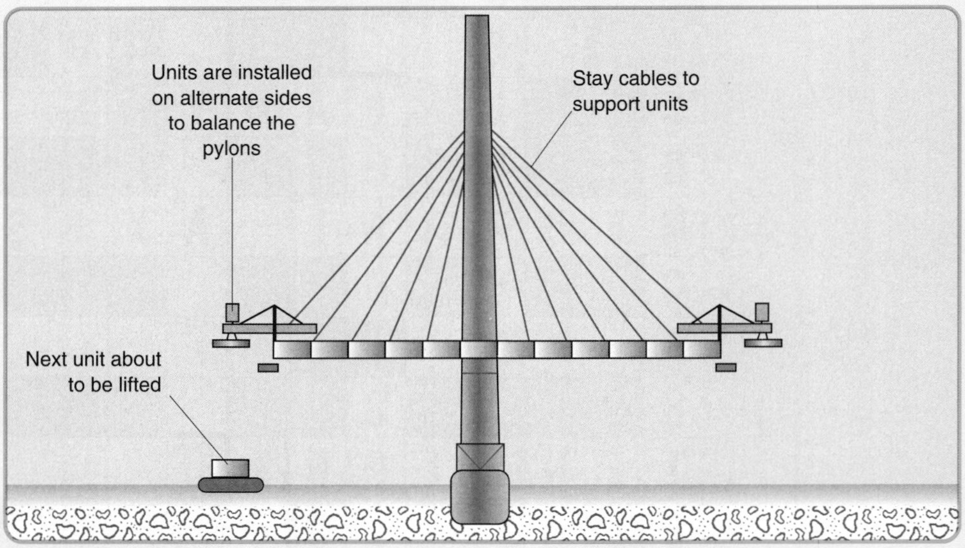

1 The overall length of the crossing is 5.2 km. The central span takes up only 456 m of this.

2 The contractor has chosen to precast all deck segments on shore. This makes full use of the limited working time available.

3 A small fleet of amphibious craft is used for the transport of the deck segments. This avoids delays due to tide changes.

4 Deck units are installed on alternate sides. This balances the pylons.

5 Casting of each pylon takes four weeks. Each pylon is cast on site up to its 137m height.

Focus The history of bridges

6 From iron bridge to suspension bridge

READING

Read and listen to the following lecture on the history of bridges given by an American professor.

In our short overview of bridge design through the centuries I would like to pick up the thread where I left it last week – the Iron Bridge at Carbrookdale. The new bridge was built as an arch bridge, because arch bridges, as we said last week, have the advantage of subjecting their structural members to compressive forces only. An interesting feature of this bridge is that it was built without any use of bolts or rivets. Instead, all the individual members were either clamped

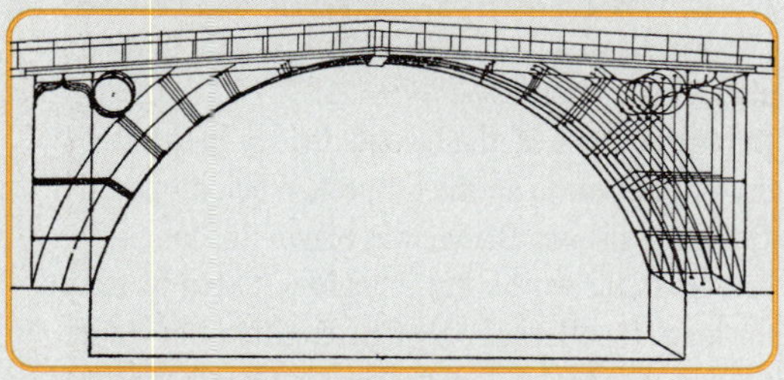

Figure 1: The Iron Bridge

together or fastened together by means of wedges. The bridge had a span of over a 100 feet and weighed approximately 400 tons.

As the science of metallurgy progressed, iron could be refined into wrought iron and steel economically and on a large scale so that steel eventually superseded the use of cast iron and, of course, that of stone and brick. Cheap steel, together with the advent of the railway age, lead to new bridge designs, in particular the Truss Bridge shown in Figure 2 and the Cantilever Bridge shown in Figure 3. The two bridges minimized the amount of material used for the weight they could carry.

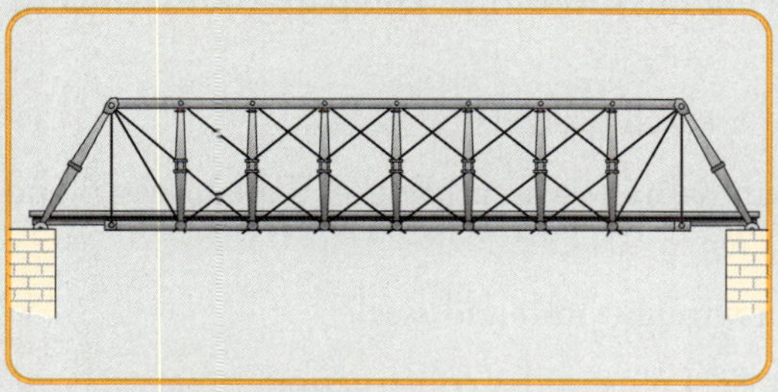

Figure 2: The Truss Bridge

Prerequisites for the construction of such bridges were not only achievements in metallurgy but also mathematical analysis. A telling example of the importance of mathematical analysis is the story of the building of the great bridge at St. Louis, USA, in the 1860s. The arches for the bridge were built in such a way that the two parts of the span were cantilevered from their respective piers towards midspan in counter-balanced stages. Temporary towers had been erected on each pier to support the half arches during construction. When the arch was about to be completed, it was realized that the gap between the two halves was too close to admit the suspended span. Mathematical calculations revealed that the gap could be increased if the temperature of the superstructure could be lowered to 44 degrees Fahrenheit, the ambient temperature being in the eighties. 45 tons of ice were carried to the site and the whole structure was packed in ice. The attempt failed, however, and finally

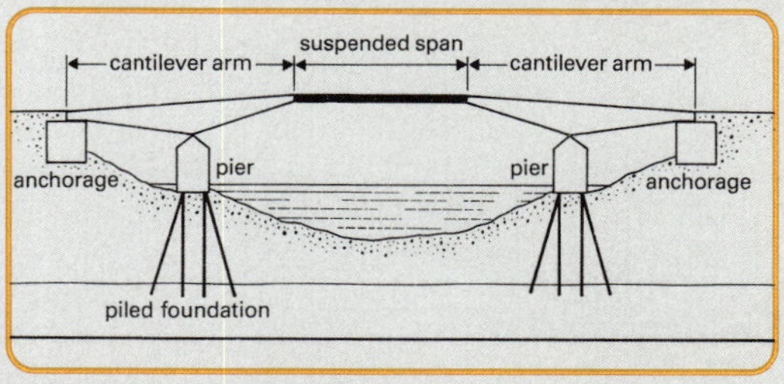

Figure 3: The Cantilever Bridge

adjustable closing members were installed which had been designed and prepared in anticipation of such an event.

Suspension bridges had always been built, but they had generally been restricted to small spans and could not carry heavy loads. The opposite came to be true towards the end of the 19th century. Ever-increasing spans became possible and the world's largest bridges were soon built as suspension bridges. The special problem of suspension bridges was, of course, that the whole weight of the bridge had to be borne by two construction members, the cables (or wires) and the piers.

The problem of producing wires of extremely high tensile strength was overcome through advances in metallurgy and production techniques. The control of the carbon content and the possibility of adding other metals to produce alloys had so much improved that galvanized steel wires could be made possessing extremely high tensile strengths. The wires would be spun together to form thick cables, on which the weight of the whole bridge could be supported.
The Golden Gate Bridge spanning the San Francisco Bay is an almost familiar sight. It is one of the longest bridges in the world and undoubtedly also one of the most elegant. But there are other remarkable suspension bridges, such as Brooklyn Bridge in New York, which in its time was the longest suspension bridge in the world.
I'll say a few words about Brooklyn Bridge in a little while, because it played an important part not only in the advancement of cable technology but also in foundation technology. But before I get to that I'd like to draw your attention to Figure 4, and in particular

Figure 4: The Suspension Bridge

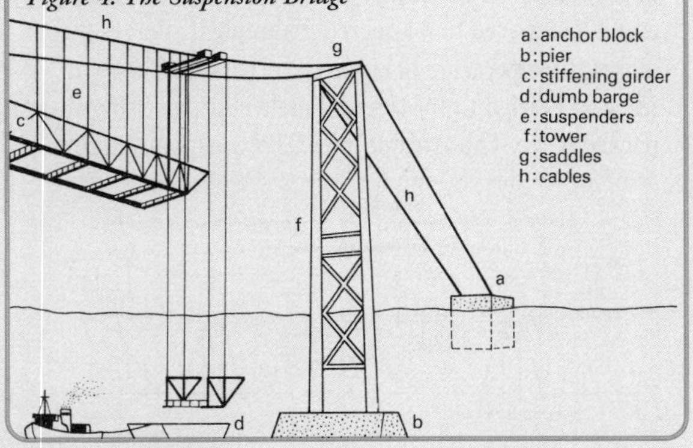

Figure 5: Tacoma Bridge

to part C of the drawing: the stiffening girder. The importance of providing a stiff deck had, of course, always been known.
But the collapse of the Tacoma Bridge in the U.S. drew attention to another aspect. When it opened in 1940, the Tacoma Bridge was one of the longest bridges in the world, but it could not keep its record for long. It collapsed only four months after it was opened. The reason for this was that winds set the bridge in a wave-like, almost galloping motion, which led to the total destruction of the bridge. It was later found that the waves were caused by inadequate aerodynamic features of the stiffening deck. Certain parts of the deck had behaved like an aeroplane's wing and were lifted up, thus causing the wave-like motion. But I've jumped ahead of myself. Let us go back to the end of the nineteenth century and look at some other trends …

7

TASK

In small groups discuss the following questions then compare your results with other groups.

1 What is the main advantage of an arch bridge?

2 Early concrete bridges were mainly built as arch bridges. Explain when and why that changed.

3 What is the difference between wrought iron and steel?

4 Make a sketch of the bridge at St. Louis under construction.

5 Why was the whole structure of the bridge packed in ice and not only the suspended span?

6 Explain the reason for the collapse of the Tacoma bridge in your own words.

7 Explain the aerodynamic features of an airplane's wing.

MODULE 4b

A television tower

Focus Illustrations in technical texts

1

STUDY SKILL

Illustrations as an aid to understanding

Technical texts are often accompanied by drawings or pictures which illustrate the text. These illustrations provide useful information on different levels.
On one level they give general information and tell you, for example, what the text is about, what processes are explained in it, etc. It is therefore useful to study the illustrations before you read the text. On another level illustrations help with vocabulary. Often they are labelled and thus provide the names of certain parts. But unlabelled illustrations can also help, because the drawing or picture provides visual information that may explain certain words in the text.

- In section 2 you can practise using the information provided by illustrations.

2

TASK

Before you read the text in section 3 about the construction of a telecommunications tower in Barcelona, look at the drawing (which accompanied the original) and answer the following questions.

1 In what chronological order do you expect the following events to be described in the text?

- **a** tying the wires to the ground
- **b** stabilizing the shaft
- **c** assembling the superstructure
- **d** construction of the shaft
- **e** pulling up the superstructure
- **f** extending the telescopic mast.

2 Which parts (1 – 5) of the drawing should be labelled as follows?

- **a** shaft/core
- **b** superstructure
- **c** temporary wires
- **d** permanent wires
- **e** telescopic mast

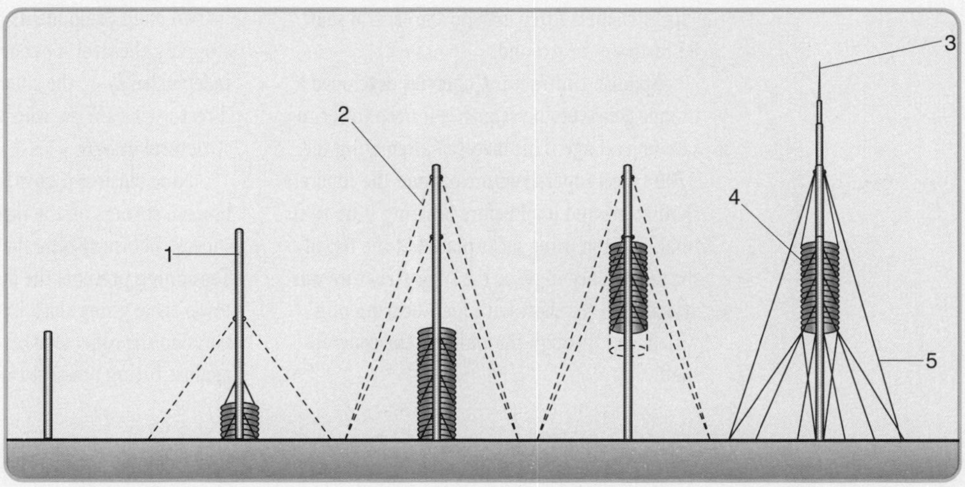

FOCUS Authentic reading

3 Raising the tower

READING

Read the text and check your answers to the questions in section 2.

DISH UP

British structural engineers played a vital role in ensuring that television pictures of the 1992 Olympics in Barcelona reached viewers around the world. The Torra de Collserola telecommunications tower, designed by consultant Ove Arup & Partners with architect Sir Norman Foster, cost £12.5 million to build and started transmitting in January 1992.

Construction of the tower (a joint project by Spain's main national television station, a local satellite TV company and the national telecommunications company) was part of a major works programme launched in the 1980s to prepare Barcelona for the Olympics.

Work in the city also involved construction of an Olympic village, increasing the capacity of the Montjuic athletics stadium from 20,000 people to 70,000, and construction of a spectacular steel and glass-covered domed volleyball stadium.

Torra de Collserola comprises a 205 m slipformed reinforced concrete shaft supporting an 83 m telescopic steel radio mast. Satellite dishes are slung from a 13-storey steel-framed superstructure fitted around the central shaft 85 m above the ground.

Spanish contractor Cubertias developed a unique construction sequence for the tower to save on cranage. This involved assembling the 2700 t steel superstructure around the concrete shaft at ground level before hoisting it up to its final position using jacks placed at the top of the shaft. Once in place the superstructure was attached to the shaft with load-bearing pins which pass through the walls of the concrete shaft.

Cubertias used temporary guys to hold the core rigid while the steel superstructure was lifted up. These were later replaced by permanent tensioned cables attached to the superstructure.

Jacks were also used to lift the tower's steel mast to the top of the concrete core. The mast comprises three telescopic sections which were contracted and placed inside the tower core when it was still under construction. Swiss jacking subcontractor VSL later raised the mast to the top of the core before extending it to its full height.

Despite its slenderness, the tower has been designed to be as rigid as a conventional high-rise building so that swaying in high winds is kept to a minimum. Rigidity was a major factor in the tower's design. Excessive sway caused by strong winds would disrupt signals bouncing from the satellite dishes, forcing the tower to shut down. Maximum deflection movements of just 85 mm are allowed 152 m up the tower when wind speeds reach 28 m/s.

The tower was stiffened by tying the steel superstructure to the concrete shaft. This made its two main components behave as one, stopping the steel superstructure from moving independently of the concrete core, says Arup director Chris Wise, who worked on the structural designs.

Nine tensioned guys fanning out from the bottom corners of the three-sided superstructure also tie the tower to the ground. Tensioning prevents the guys downwind of the tower from going slack under wind loadings. Guys on the other side of the tower also pull against tilting pressure exerted by wind.

4
TASK

Decide whether the following statements are true or false. Correct the ones that you think are false.

1 As part of the preparations for the Olympics in Barcelona the national telecommunications company had a television tower built.
2 The tower measures a total of 205 metres in height.
3 The superstructure was built on the ground and later raised into position.
4 The central shaft consists of three sections which are held together by cables.
5 The telescopic mast is fastened to the ground by means of tensioned cables.
6 The design of the tower makes it as stable as a traditional high-rise building.

5
VOCABULARY

Which words in the text mean the following?

1 essential to the existence, or operation of something (adj)
2 a project which was carried out by several companies (n)
3 impressive or extraordinary (adj)
4 unusual, special (adj)
5 to pull up (v)
6 stiff, not bending or yielding (adj)
7 not temporary (adj)
8 to consist of (v)
9 to make smaller or shorter (v)
10 to disturb, to cause disorder (v)
11 element of, part of (n)
12 (to cause) to move sideways (v)

6
STUDY SKILL

Writing a summary

To summarise a text follow the following procedure.

1 Read the text carefully several times, making sure you have fully understood it.
2 Close the magazine or book and try to think of a limited number of key ideas. Write them down in not more than one sentence each. (The number of ideas will depend on the length of the text.)
3 Read the text again.
4 Check what you wrote in step 2, adding any ideas that you think are also important, and/or crossing out any you think are not really important.
5 Close the magazine or book again and put your ideas into the order you want to include them.
6 Write a first draft of your summary. It is better not to read the article while you are writing the summary. Leave out specifics that you are not quite sure about.
7 Read the article again. Read your summary (if necessary add the specifics) and finalise it. Make sure you do not go beyond the main ideas you decided on in steps 2 and 4.

- Now practise the above procedure with the text in section 3.

MODULE 5a

Providing and recycling

Focus Water treatment

TASK 1 The illustration depicts a water treatment plant near London on the river Thames. What do you think happens at the stages indicated by the numbers 1 – 7?

2 Thames Water

READING

The following leaflet describes the water treatment plant in the illustration. Complete the text with (correct forms of the) verbs and prepositions from the following list.

to work on • to go to • to turn to • to take from (2x) • to supply to • to feed into • to force through • to drain down • to pass through • to sink through • to feed on • to come from • to wash away • to leave behind

We … (1) water for a hundred uses. We take it for granted, because it costs very little, but where does it come from? It is … (2) us by clouds, rivers, boreholes and the Regional Water Authorities.

Most of the region's water is … (3) the network of rivers and streams, all of which eventually … (4) the Thames.

Water is taken from rivers at intake points with screens to catch floating debris and is lifted by water pumps to a storage reservoir.

The total storage capacity of the Thames region reservoirs ensures that when water is plentiful it is conserved as a reserve in case of drought. And they are an important preliminary stage in the natural purification of water on its way to the tap. The comparative stillness of the water allows the bigger impurities to settle, and the large surface area allows oxygen in the air to … (5) the other impurities.

From the storage reservoirs the water is taken for treatment. Methods vary but the effect is the same. First comes primary filtration. One method of primary filtration uses rapid filter beds.

The water … (6) a layer of coarse sand at a rate of twenty feet an hour, … (7) the larger suspended matter. It finally passes through a layer of shingle into perforated pipes, which carry it away. Every day the filter beds are … (8) to the last foot or so of water and then compressed air is … (9) the perforated pipes to loosen the dirt, which is … (10) by a reverse, upward flow of clean water.

After this comes secondary filtration. The principle is the same as in primary filtration, but in these beds the sand is finer. Minute natural organisms in the sand … (11) bacteria in the water and help purify it further. So beneficial is slow sand filtration that it has been said to be the most important single method of water treatment ever developed for the prevention of disease.

Some of the region's water is … (12) chalk wells and boreholes in London and elsewhere. It is much purer than river water and does not require filtration.

All the water … (13) a covered tank where the action of chlorine kills any remaining bacteria. The water is then safe to drink and ready for pumping into the mains.

Trunk mains carry it underground or overground and whatever is not used en route goes into water towers or service reservoirs. Sometimes these are built unobtrusively under parks, golf courses or football pitches. The service reservoirs have to be on high ground. This is so that during the day, if demand for water is high and pressure in the main falls, water from the pumping station can be supplemented by gravity feed from the service reservoirs. At night, when the demand is low, the reservoir can be pumped full again.

From the service reservoirs, a network of mains … (14) the consumers. And from these mains, a service pipe brings water direct to the household water tank. Every day, water samples are tested in Thames Water's quality control laboratories. Samples … (15) all stages of water treatment. Thames Water is very conscious of its responsibilities for keeping a constant check on pollution throughout the whole system.

FOCUS Sorting techniques in recycling

3 A plant manager explains the sorting system in a recycling plant

LISTENING

Listen to the plant manager and answer the following questions.

1 What kind of waste is processed in the plant?

2 Is the process fully automatic?

3 Why are some plastic articles shredded?

- Read the statements below, then listen again. Take notes that will help you say whether the following sequence of events is true or false and if any steps are missing.

1 The recyclables are collected in the community.

2 They are put on an electromagnetic conveyor belt.

3 Ferrous materials are sorted out.

4 Aluminium is sorted out.

5 Glass is sorted out.

6 Plastic parts are shredded, washed and ground.

7 In the hydrocyclone the plastic is separated.

8 The plastic is turned into pellets and packed.

4 VOCABULARY

Each line contains one word which does not quite fit. Which one is it: a, b, c or d?

1	a around	b vertical	c upwards	d inclined
2	a to exit	b to spin	c to travel	d to dry
3	a fairly	b manually	c absolutely	d easily
4	a the flakes	b the pellets	c the recyclables	d the strips
5	a to grind	b to shred	c to break	d to sort
6	a the weight	b the density	c the process	d the magnetism

PHRASES

Describing processes

The first/second/third stage is …	First(ly), … . Second(ly), … . Third(ly), …
Subsequently/After that …	Now/Then …
After/Before …	While/When …
Once/During/As …	

5 WRITING

Use the notes you made in section 3 to write a description of the sorting system in the recycling plant.

FOCUS Waste incineration

6 A waste-incineration plant

READING

The sorting and recycling of municipal waste is very costly, and alternative methods have been developed. These include modern waste-incineration plants in which the calorific value of the waste is used to generate electricity, and the exhaust gas is carefully treated.

The jumbled paragraphs below describe the stages of the waste-incineration process shown in the illustration. Put the paragraphs into the correct order. The first, sixth, tenth and fourteenth steps are already numbered.

SHREDDING AND BURNING TRASH FOR ENERGY

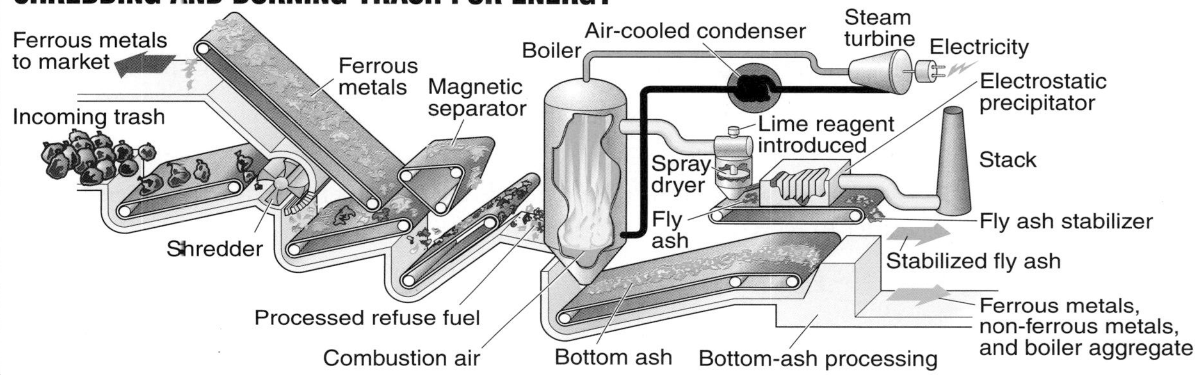

☐ This picks up all the ferromagnetic parts and subsequently drops them onto a conveyor belt which takes them to the bin for ferrous metals.

1 The illustration shows a waste-incineration plant that produces more than 50 megawatts from trash.

☐ The grate temperature is kept below the melting point of glass so that the glass, together with other non-combustibles such as aluminium, becomes part of the boiler aggregate which is sifted from the bottom ash.

☐ After the gas has passed the scrubber, it passes through the electrostatic precipitator, where the particles are first charged electrostatically so that they then attach themselves to oppositely charged plates.

☐ From the plates they are mechanically removed and leave the plant.

6 The remaining trash fragments are now blown into the boiler.

☐ The trash for this plant, which is situated in Rochester, Massachusetts, is collected in 32 communities.

14 After the steam has left the turbine it is condensed and the water returns to the boiler to complete the cycle.

☐ The heat produced in the boiler is used to generate steam which drives the turbine.

10 The flue-gas treatment involves two steps. First the gas passes through an acid gas scrubber where lime reacts with the acid gas to neutralize it.

☐ The trash arrives at the plant by rail where the cars tip the trash onto a conveyor belt which takes it to the hammermill shredders.

☐ The exhaust gas, also called the flue exhaust, must be treated before it can be released into the atmosphere.

☐ The trash is shredded and falls onto another conveyor belt which passes a magnetic separator.

☐ The temperature in the boiler is well above 2,000 degrees Fahrenheit so that all the combustible objects ignite immediately while the heavier non-combustible ones fall to the grate.

FOCUS Recycling plastic

7 TASK

You are going to read and listen to a lecture text about the properties, the production and the recyclability of plastics. Before you start the text, get together in small groups and read the statements below. Decide which of the sentences are true/untrue according to what you know about plastics. Then go on to section 8, read and listen and check your choices.

1 There are two basic kinds of man-made materials: plastic and polymer.

2 In general plastic is non-corrosive.

3 Plastic consists of long chains of molecules.

4 There are two different kinds of polymers: thermoplastic polymers and thermosetting polymers.

5 Thermoplastic polymers can be reheated and reformed several times.

6 Thermosetting polymers harden while they are still hot. They cannot be reheated and reshaped.

7 Because they cannot be reheated and reshaped, thermosetting polymers are easier to recycle. They can be burnt.

8 Plastic products are always homogeneous. They consist of one type of plastic only.

8 READING Polymers

POLYMERS

Now, we all know that plastic products are used for almost any application, from shopping bags to car parts. This is because they are cheap, light, non-corrosive – in fact they can be given almost any desirable property. And because this is so, they have become almost omnipresent. Plastics, or polymers, which is the more scientific term for them, are everywhere. For the recycling industry, however, they can be a problem. To analyse this problem we have to look at how polymers behave during production. Here it is necessary to distinguish between two basic types of polymers, the thermosetting polymers on the one hand and the thermoplastic polymers on the other hand. Let's look at the second type, the thermoplastic polymers first. Thermoplastic polymers are made up of very long chains of molecules with very strong internal forces that keep the atoms in the molecular structure. Also, the forces that attach one molecule to the next in line are also very strong. Not strong, however, are the

POLYMERS

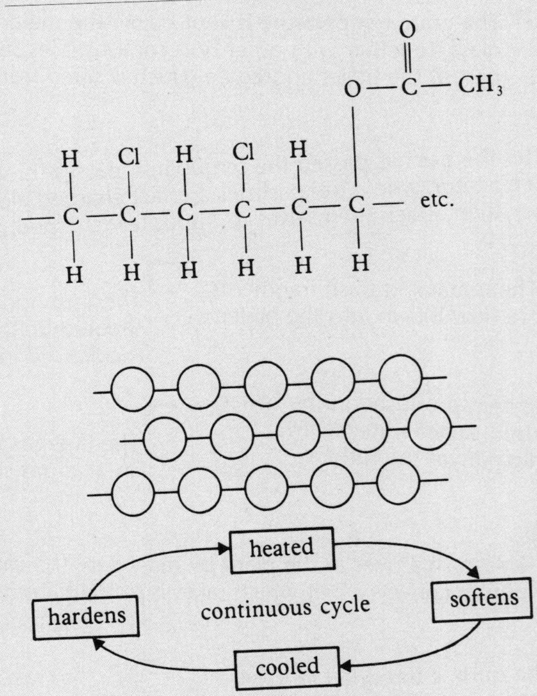

Illustration A: Thermoplastic polymers

POLYMERS

attractive forces between adjacent chains. Now, what does this mean in terms of plastic production and plastic recycling? What happens is this. When you heat up a polymer with relatively weak attractive forces between adjacent molecule chains, those chains slide along one another and the material becomes soft. It will remain soft until the heat becomes either too high and the material begins to burn or until the temperature drops and the material hardens. While the polymer is in the soft state it can be shaped into different forms. The main advantage of these thermoplastic polymers is that the process of heating and forming with subsequent cooling can be repeated very many times without a high degree of deterioration of the material. In other words, for the recycling industry this means that such materials can be very easily recycled because the softening and hardening can be repeated many times without involving any great chemical change.

Now, let's come to the thermosetting plastics. These behave very differently. Subjected to heat for the first time they soften and can then be shaped into the desired forms. But instead of remaining soft while they are hot, they already harden in the hot state and they keep their shape and mechanical properties while they are heated further and when they cool down. Reheating will not change either their shape or their mechanical properties. In other words, these materials cannot be recycled by way of reheating them and giving them a new shape. This is because these plastics are made up of lines of molecules which are heavily cross-linked.

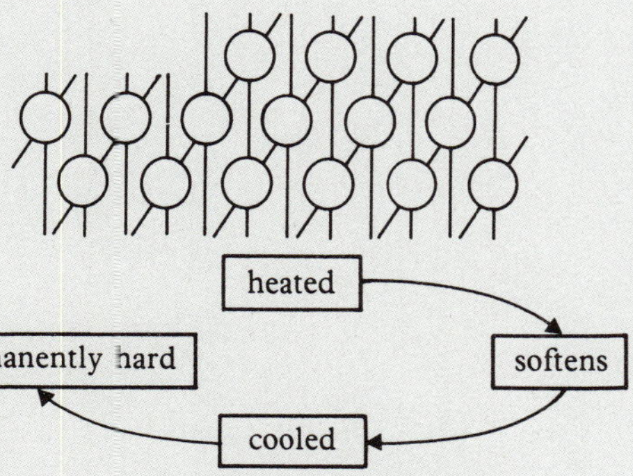

Illustration B: Thermosetting polymers

POLYMERS

This results in a rigid molecular structure with very strong bonds between the molecules so that this structure cannot be altered; it can only be destroyed by burning. On the positive side such materials are usually much stronger than the thermoplastic materials. And from a production point of view they have certain advantages, but I won't go into that now. For the recycler, however, they are bad news, as they can't really be recycled. Recently, however, new technologies have begun to be developed and it may very well be that some time in the not too distant future we'll have decomposition technologies that will break them down into useful products such as oil or gas.

Maybe a last word about thermoplastics. I have, perhaps, given you a slightly oversimplified view of the recyclability of thermoplastics, for, strictly speaking, what I said applies only to homogeneous plastic waste. The problem is, however, that many plastic articles are not homogeneous and consist of different kinds of polymers which makes matters more complicated.

Now I think I should stop here. We have a leaflet summarizing the main points. You may want to have a look at it later and if there is anything you don't understand, don't hesitate to ask me.

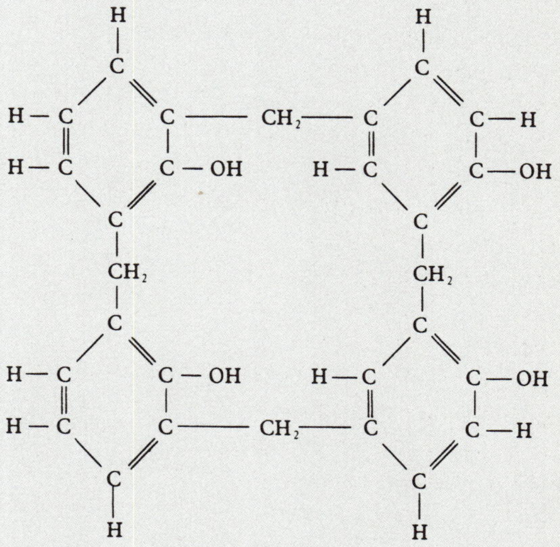

9 TASK

Answer the following questions on the text in section 8.

1 What are the main advantages of plastics?

2 What is the chemical difference between thermosetting and thermoplastic polymers?

3 Explain the difference between thermosetting and thermoplastic polymers in terms of recyclability.

4 What alternative may there be in the future to burning thermosetting polymers?

5 What problems do thermoplastic polymers cause in terms of recycling?

10 PRACTICE

Make true sentences from the following sentence switchboard. (You may want to look back at the phrases used to give reasons on p. 26).

Due to	the weak forces between adjacent molecular chains	plastics can be used in very many different applications.
On account of	their diverse properties	they are usually burnt.
Because (of)	reheating does not change their mechanical properties	they have become a symbol of the throw-away society.
Owing to	the strong forces between the molecules	thermoplastic polymers can be reheated and reshaped.
As	the low price of plastics	reheating does not change the chemical properties of thermosetting polymers.
	the difficulty in recycling thermosetting polymers	thermosetting polymers are difficult to recycle.

MODULE 5b

The tyre problem

Focus Authentic reading

1

READING

Tyres that burn, tyres that dissolve

Split into two groups. Group A reads the article *Tyres in inferno yield oil and gas*, Group B reads the article *Old tyres don't die, they just dissolve away.* Each group thoroughly prepares its article, i.e. looks up relevant vocabulary etc. When both groups are finished, Group A reads out "their" text, taking turns from paragraph to paragraph. Group B also reads the text and can/should ask questions relating to vocabulary, grammar and meaning. These are then answered by Group A. Later, follow the same procedure with the second text.

Tyres in inferno yield oil and gas

A cheap way of turning the world's ever-rising tyre mountains into gas, oil and other products has been developed in Britain. Engineers from AEA Technology and the company Herbert Beven have devised a system which decomposes tyres at high temperature. They claim it is cleaner and more economic than incineration.

A pilot plant has been tested at an AEA Technology site near London. Tyres are heated up to 1080°C in a high-temperature kiln, which breaks them down into two fuels, gas and oil, as well as "char", a substance rich in high-quality carbon and steel. The kiln is fitted with a condenser to capture oils in the exhaust and a gas scrubber to clean up its gas emissions.

Fires in tyre dumps in Powys, Wales, and most dramatically at Hagersville, Canada, have highlighted the environmental danger of dumps. The Canadian fire released large quantities of dioxin and aromatic hydrocarbons into the atmosphere, and oil into groundwater. The US alone is estimated to have 3 billion tyres in dumps. Previous large-scale plants designed to decompose tyres with heat were a technical and financial failure.

Now the first commercial Multi-Purpose Disposer (MPD) has been sold to a company in the US, which will then manufacture and market the British design in North America.

Each new plant can handle around 300 tyres – about 1000 kilograms of material – in less than nine hours. Twenty plants can be linked up to deal with 14000 tonnes of tyres a year. Each tonne of European tyres produces up to 220 kilograms of aromatic oil, 240 kilograms of gas, 420 kilograms of carbon and 160 kilograms of steel.

The char is of sufficiently high quality and calorific value that it can be used either in filters or as fuel in its own right. The oil and gas can be piped straight into a nearby combined heat and power plant.

Herbert Beven says that MPD currently satisfies all European emission standards for pollutants and could be licensed anywhere in the US except California, home of the toughest air pollution regulations in the world. "But with some relatively simple refinements we will be able to meet those too."

The system produces lower emissions of nitrogen oxide and sulphur dioxide than incinerators designed to burn tyres and produce electricity, say the engineers. Britain's first such plant, designed to produce up to 25 megawatts of electricity for the national grid, is being built at Wolverhampton. The scheme is 18 months behind schedule and will not now be ready until the end of the next year. Once operational it should burn some 12 million tyres a year.

Old tyres don't die, they just dissolve away

A Japanese research group claims to have found a way of banishing the environmental headache of discarded tyres – decomposing them into recyclable components. The need worldwide is enormous: in the US alone, more than 230 million tonnes of used tyres pile up every year. About 80 percent end up in landfills, while most of the rest are either dumped or burnt.

The researchers use a combination of "supercritical" water, which is at such a high temperature and pressure that its liquid and gas states merge, and sodium hydroxide. The organic components of the tyres can then be reused in other products.

The technique was developed over the past few years by a research team led by Toshinari Tennoh of the Nishikawa Rubber Company in Hiroshima. The tyres are cut into pieces a few centimetres across and then heated to 400° C at a pressure of more than 40 atmospheres in a one-molar (about 4 per cent by weight) solution of sodium hydroxide. Under these conditions, water can dissolve organic compounds. After 15 minutes the rubber in the tyres decomposes into an oily mixture of long-chain hydrocarbons. This mixture is separated from the steel and textiles within the tyres, which are then recovered for recycling.

Nishikawa's process can also separate and recover the sulphur added to most tyres in the vulcanisation process that hardens them. The sulphur forms crosslinks between the rubber polymer chains, making them more rigid.

As the rubber decomposes during the recycling process, these sulphur crosslinks break and the sulphur is converted to hydrogen sulphide. This gas is then passed over zinc oxide pellets, forming zinc sulphide. No organic sulphur compounds have been detected in the hydrocarbons recovered from the process for recycling.

The advantage of this process is that it goes further than any other in devulcanising the rubber and decomposing it into smaller organic molecules, so more products can be made from them. Other processes are less efficient at removing the sulphur used in vulcanisation.

2 VOCABULARY

Complete the following crossword. Except for four clues, all the words occur in the two texts above.

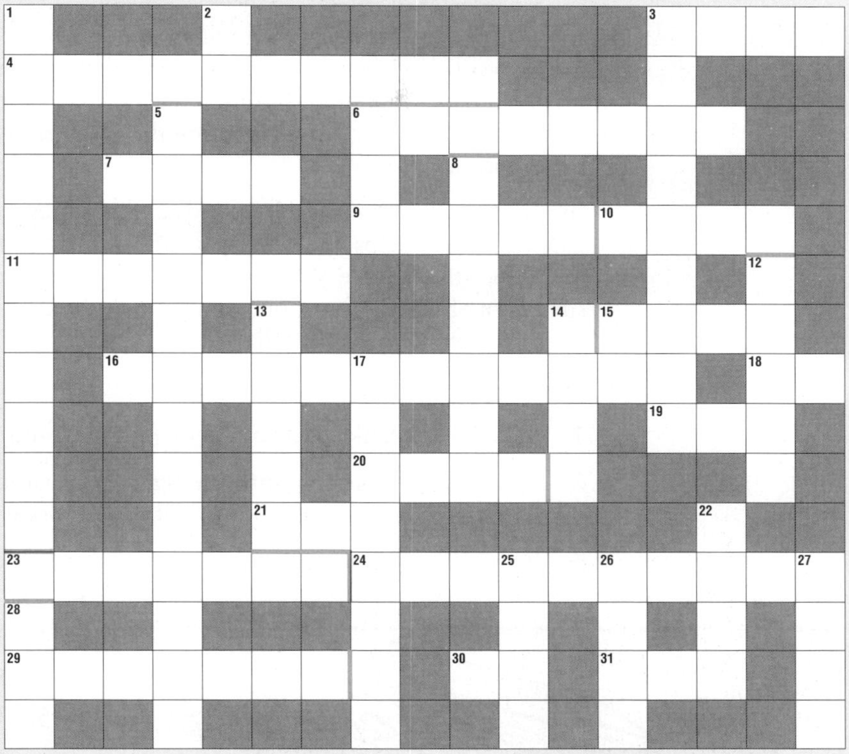

Across

3 The British word meaning mail in American English.
4 The process of making something purer or finer.
6 The element created when two or more atoms combine.
7 The (frequently illegal) graveyard for tyres.
9 If Fe combines with C, the new alloy is tougher and harder than Fe.
10 A word often employed to mean "very clean".
11 A verb meaning "to throw away".
15 Another verb for "being unsuccessful".
16 A synonym for "burning" (noun).
18 A preposition sometimes mistaken to mean "on".
19 A unit of measurement.
20 A single thing or piece. A list usually consists of many of them.
21 A prefix which is similar in meaning to the prefix "un-".
23 Another word for "to catch".
24 A synonym for "enough".
29 This word is a key-word in recycling because it is used to mean that one can get back something.
30 A prefix which turns things around.
31 This word combines and connects.

Down

1 A very important natural resource which used to be in abundant supply in most European and North American countries but is now threatened not only by pollution but even by shortage.
2 A preposition often taken to mean "into". But there is a difference.
3 The general name given to the agent which destroys the purity of something else.
5 A chemical process which gives tyres the necessary strength and elasticity.
6 The prefix which indicates that something is done improperly.
8 To set free or liberate or the act of setting free or liberating.
12 Depending on the context this word can mean either a kind of flower or a place where something is produced.
13 An adjective often used as a synonym for "inflexible".
14 A kind of oven used in production processes.
17 Another word for the "escape of gas".
22 A non-ferrous metal which is very malleable, not elastic and very dense.
25 The resource on which the prosperity of industrialized nations depends.
26 A solid end product of the decomposition of tyres which is rich in carbon.
27 Two or multiples of two are needed in transportation, unless it is a spare one.
28 A prefix meaning "before" or "first".

FOCUS Giving a presentation

3 Mind mapping

STUDY SKILL

Engineers and technicians, as well as students, are often required to give a short presentation. The best presentations are usually those that are not read from a prepared paper, but are given more or less freely with the help of just a few notes or diagrams. Mind mapping is a technique that helps you prepare and give a presentation.

In mind mapping you use keywords and pictures to create a "map" of your talk. The technique helps you focus your ideas during the preparation phase, and also to remember the key ideas during the actual presentation. The mind map is also a summary of the presentation, so it can also be used on the overhead projector to help listeners follow your talk.

Here are some guidelines for mind mapping.

1 Take a few blank sheets of A3 or A4 paper and lots of coloured felt-tip pens.

2 Relax yourself, getting into the right frame of mind for new, creative thought, and forget your belief in your inability to draw.

3 Select your key word(s) or image, and write or draw it in a box in the centre of your page.

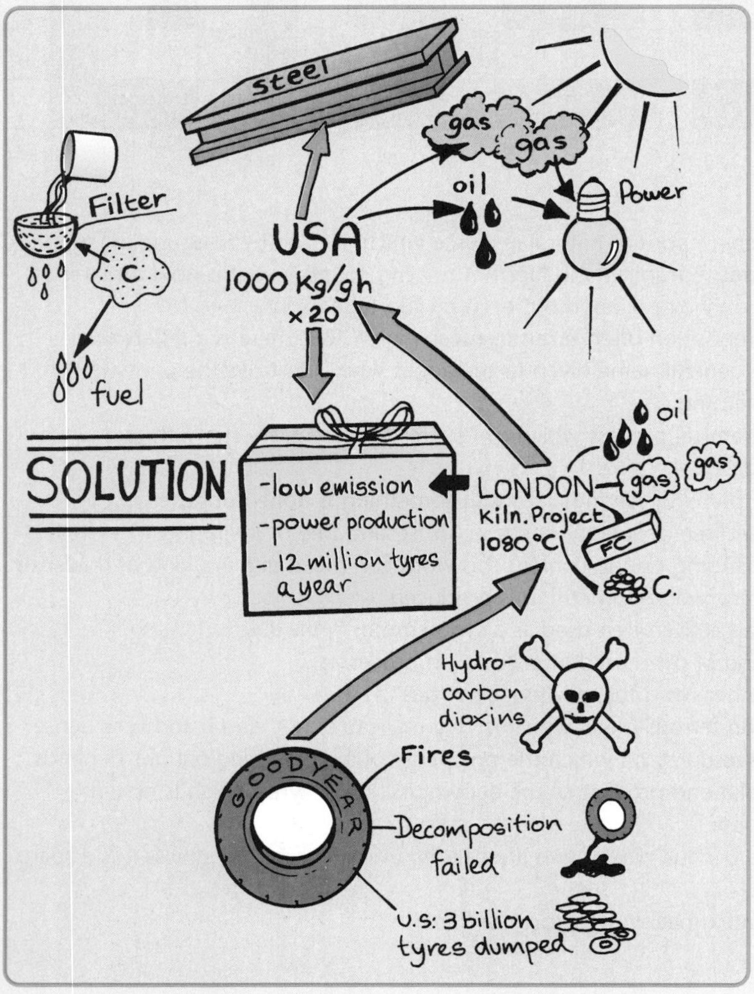

4 Draw any ideas related to this central theme next to the box.

5 Use the pens to differentiate your ideas, and draw links between the ideas.

6 Condense your thoughts to one word per line.

7 Use images instead of words, or in addition to the words, wherever possible. The more imaginative and colourful the mind map, the better you remember your thoughts and ideas.

8 When you have finished, edit and regroup your notes on a fresh sheet in order to produce your final, master mind map.

The mind map on the left was prepared for a short presentation on the British/American solution to decomposing tyres *(Tyres in inferno …)* in section 1.

Now draw your own map and give a presentation of the Japanese solution *(Old tyres don't die …)*.

MODULE 6a

Waves and systems

Focus Waves

TASK 1

The definitions below are not in their right place. Find the correct definitions (A to L) for drawings and terms 1 to 11.

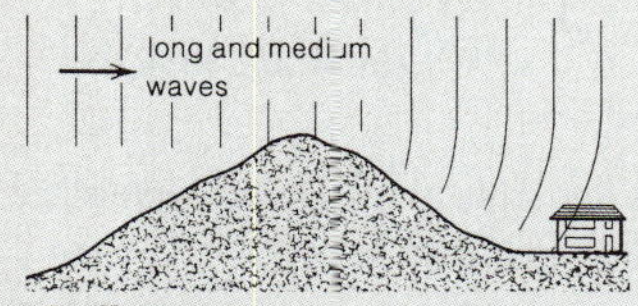

1 Diffraction

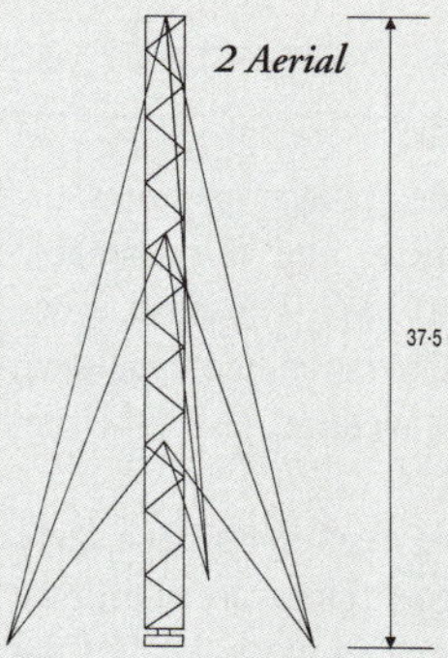

2 Aerial

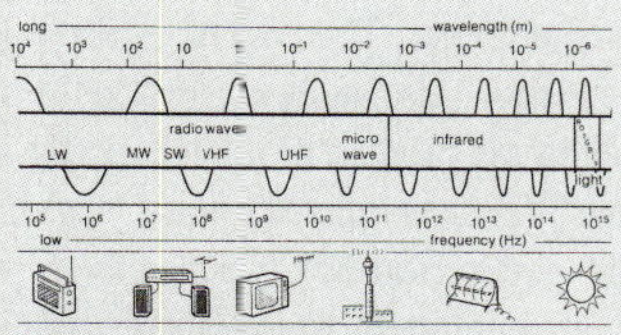

3 Electromagnetic waves

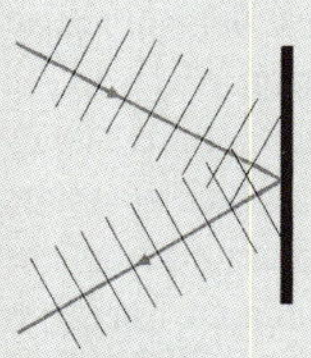

4 Reflection

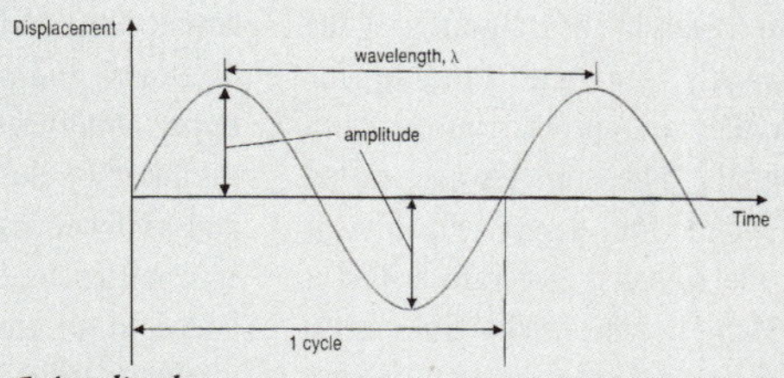

5 Amplitude

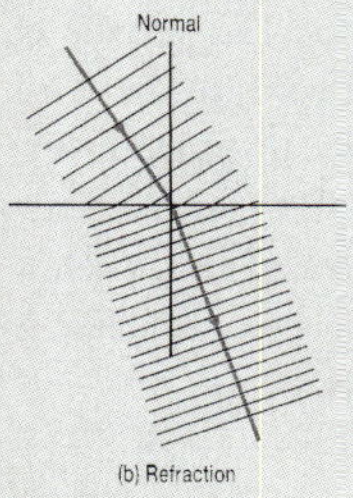

6 Refraction

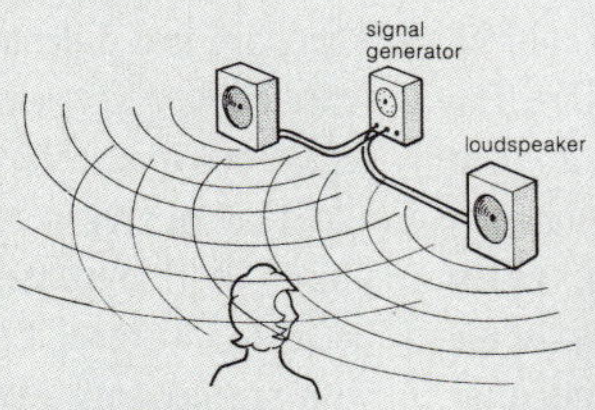

7 Interference

8 Laser

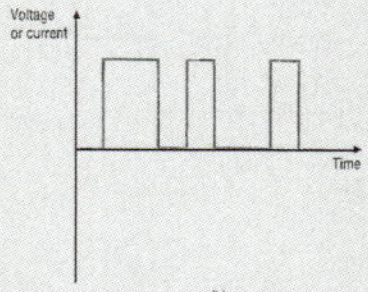

9 Digital signals

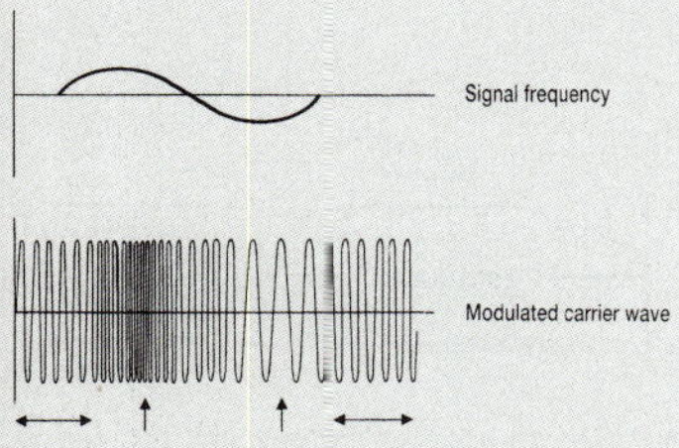

10 Carrier wave

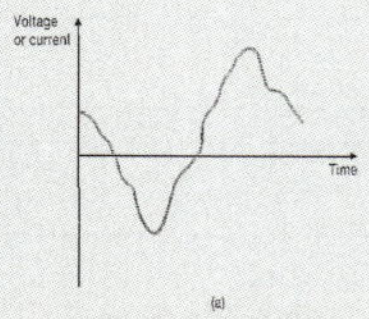

11 Analogue signals

A The family of rays and waves that includes radio waves, microwaves, X-rays etc. All consist of vibrating electric and magnetic fields and travel at 300,000 kilometres per second. All the waves and rays differ only in their wavelength or frequency. Except for gamma rays, they are all generated by accelerating electrons.

B They have only fixed values which are transmitted as a train of pulses.

C The amount of energy in a wave. It is equal to the change in energy (for example, pressure in sound waves) that takes place as one complete wave passes. Graphically speaking, it is the distance between the middle and the top of a wave.

D The bending of waves that occurs as they pass through an opening or over hills. The angle of bending depends on the wavelength.

E The radio signal that is broadcast at a particular frequency or wavelength and which is modulated to carry a sound or picture signal.

F The bending of a wave or ray that occurs as it passes from one medium or substance into another, for example from air into glass.

G The reversal of direction that occurs when a wave or ray bounces off a surface.

H They vary constantly with time. In telecommunication systems they are used for local telephone networks and the majority of radio and microwave transmissions – although this may change in the near future. They require amplification at intervals.

J The part of a radio transmitter or receiver that sends out or picks up radio waves. Also referred to as an antenna.

K A device that produces a narrow beam of very bright light or infrared rays, in which all the waves have exactly the same frequency, are in phase and move exactly together.

L The effects produced when two waves or rays meet. The combined wave has a different frequency or amplitude often causing sound (or picture) distortions.

FOCUS A basic telecommunication system

2 Basic devices

READING

The following text explains some basics of telecommunication systems.

Read it and name the devices labelled A – E in Figure 2.

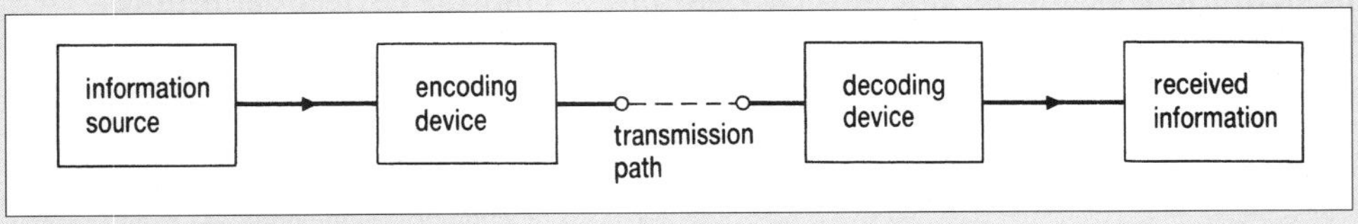

Figure 1: A basic telecommunication system

Although individual telecommunication systems do not appear to have that much in common they can be shown to be composed of similar sub-systems. The actual sub-systems may be very different in construction but perform similar functions.

A non-electrical telecommunication will contain the sub-system illustrated in Figure 1. The information from the source has first to be encoded into a form suitable for transmission. The encoded signals are then transmitted to the receiver, where the signals are decoded to retrieve the original information. For example, in sending messages by using drums, the information has first to be encoded by the sender into a series of sounds. The sound waves travel through the air (the transmission path) to the receiver, whose ears and then brain decode the sounds to recover the original signal.

The sub-system can also be applied to simple electrical telecommunication systems, but the encoder has to be a transducer. A transducer is a device that changes one form of energy into another. In the telephone system, the transducers are the microphone and the earpiece. The transmitting microphone changes sound waves into varying electrical currents, which are sent by wires to the receiving earpiece, and this changes the varying currents back into sound waves. In other systems the transmitted signals may be produced from pictures, text or computer data.

With the present electronic systems there are more sub-systems. These are illustrated in Figure 2, although a particular system may not use all the sub-systems shown. The information source has first to be changed into electrical signals by the transducer. If the resulting signals are not suitable for transmission they are encoded into a more suitable signal. The process is called modulation, and the device capable of modulating the signal is generally called a modulator. Many systems need a transmitter for getting the signal into the transmission path, and a receiver at the other end to extract the signal. The transmission path could be wire, coaxial cable, air or an optical fibre. The received signal is demodulated and then passed through the appropriate transducer to recover the original information. Often amplifiers are added to the system to boost the signals before they are transmitted and after they are received and, if necessary, also in the transmission path. Amplifiers are represented as triangles in Figure 2.

Amplifiers are necessary because some energy is always lost and so the power carried by the signal decreases. This loss of power is called attenuation. Also, noise is added to the signal, and this can produce distortions of the signal when it is demodulated. Noise refers to any random electrical energy added to the signal being transmitted. Noise can be classified as natural or artificial. Natural noise includes cosmic noise and atmospheric noise produced by storms and other electrical effects in the upper atmosphere. Artificial noise is mainly produced by electrical devices such as motors and switches. As the signal travels from the transmitter to the receiver, noise energy is added to an attenuating signal. If nothing was done to boost the transmitted signal, it would eventually disappear amidst the noise.

There are two different kinds of amplifiers: repeaters and regenerators. The purpose of repeaters is to increase the power of the signal and that of regenerators to produce new pulses. With repeaters, the noise already added to the signal is also amplified, whereas with regenerators a completely new pulse is generated that is noise-free.

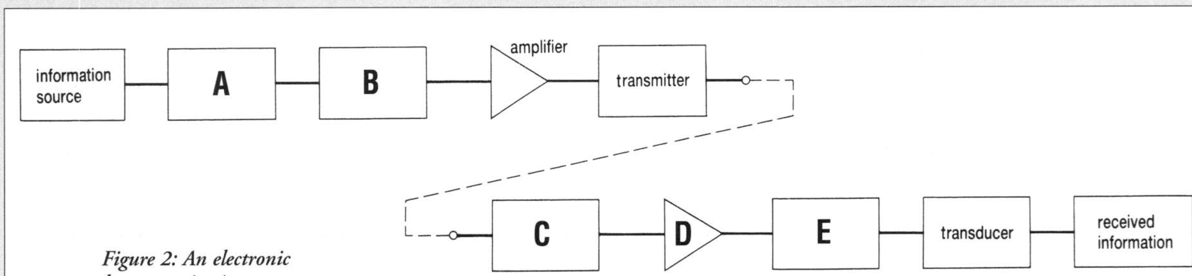

Figure 2: An electronic telecommunication system

PRINCIPLES OF TELECOMMUNICATIONS 15

3 PRACTICE

Finish the following sentences about the text above.

1 A transmitter serves to …
2 The purpose of a receiver is to …
3 Modulation and demodulation are required if …
4 The function of an amplifier in telecommunication systems is …
5 The term attenuation refers to …
6 The difference between repeaters and regenerators is …
7 A transducer is a device that …

4 PRACTICE

Make true and grammatically correct sentences from the sentence switchboard below.

Attenuation	results in/from	a loss of wave energy.
Wave interference	is due to	sound distortions.
Signal amplification	is caused by	improved transmission of information.
Artificial noise	causes	a stronger signal.
The use of digital signals	leads to	powerful electrical appliances.
A new, noise-free pulse	is the effect of	using regenerators in the transmission path.

GRAMMAR

Infinitive and -ing form

1	Individual telecommunication systems **appear to have** little in common.
2	He **suggested taking** a break.
3	Most rays are generated **by accelerating** electrons.
4	A device capable **of encoding** a signal into a more suitable form is called a modulator.
5	I **stopped to watch** the sports programme on TV.
6	I **stopped watching** the sports programme on TV.

- If a verb is followed by another verb, the second verb may be in the infinitive or -ing form.
- Verbs followed by the infinitive (example: sentence 1):
  be able, afford, agree, allow, appear, arrange, ask, attempt, cause, choose, decide, determine, enable, expect, fail, forget, intend, learn, manage, promise, permit, pretend, tend, threaten.
- Verbs followed by an -ing form (sentence 2):
  admit, avoid, consider, delay, deny, enjoy, finish, imagine, involve, mind, miss, postpone, practise, regret, risk, suggest.
- If a verb follows a preposition (sentences 3 – 4) it is always followed by the -ing form.
- Some verbs (sentences 5 – 6) can be followed by the infinitive or the -ing form: stop, remember, forget, regret, try. The meaning is different. Check your grammar book for details.

5
PRACTICE

Choose the right verb for each gap, and decide whether to use the infinitive or the -ing form.

boost • change • decrease • deteriorate • have forgotten • operate • recover • send • transmit • use

1 Coaxial cables are capable of ... several hundred calls simultaneously.
2 The loss of energy in the original signal causes the strength of the signal ... .
3 To prevent the signals from ... repeaters are installed at intervals.
4 One should avoid ... strong electric motors in close proximity to telecommunication cables.
5 Microphones are used for ... sound waves into electric signals.
6 Encoding signals in a more suitable form for transmission enables us ... signals more effectively.
7 Many companies have stopped ... telexes. They prefer sending faxes instead.
8 I regret ... to fax you the directions for how to get to us.
9 Repeaters are employed ... the incoming signal (including added electronic noise) while regenerators are able ... the original signal.

GRAMMAR

Uncountable nouns

Can you give me **some information?**

- The noun *information* is uncountable. There is no plural form (~~informations~~) and it is never preceded by an indefinite article (~~an information~~).
- Words and expressions which often precede an uncountable noun are: *some*, *a lot of*, *a piece of*, *much*, *any*, etc.
- Other uncountable words in English: access, accommodation, advice, baggage, behaviour, bread, furniture, hair, knowledge, luggage, permission, progress, scenery, traffic, trouble, weather, work.
- In a dictionary uncountable words are marked [U] and countable words are marked [C].

Now choose five uncountable words as a basis for a short (humorous) story.

6 Noise-suppression systems

LISTENING

You are going to hear part of a radio programme about new technical developments. Before you listen, make sure you know the meaning of the following words. Use a dictionary if other course participants cannot help:
the power socket, to cherish something, the chat, the obstacle, to cancel.

- Now listen and decide whether the following statements are true or false.

1 Some airplane companies provide laptop computers for their 1st and business class travellers.

2 Passengers will be able to bring and view their own video tapes.

3 Commuter planes are often noisier than bigger planes.

4 SAAB's new system operates on the basis of wave interference.

5 The noise coming from the engines of the plane is cancelled before it gets into the cabins.

6 The system has proved reliable in SAAB planes operating in the USA.

MODULE 6b

Telecommunication links

Focus Authentic reading

1 Satellite and wireless networks

READING

Split into two groups. Group A reads text 1, Group B reads text 2. Prepare your texts thoroughly using dictionaries if necessary. Take notes. Once both groups feel they have fully understood the texts, participants from Group A form pairs with participants from Group B. Now explain to one another how the respective telecommunication systems operate and find answers to the following questions.

1 Are both systems designed to be used worldwide?

2 Why does Mtel's system *(Text 2)* use separate transmission and receiving stations?

3 Why can the handsets used for Mtel's system operate with very little transmitting power?

4 The Motorola system *(Text 1)* relies on the use of 66 satellites. How many satellites do you think Mtel's system requires to service the largest 500 cities in the USA?

5 What is the purpose of the relay satellite in the Motorola system?

Group A: Text 1

Go-anywhere cellular phone

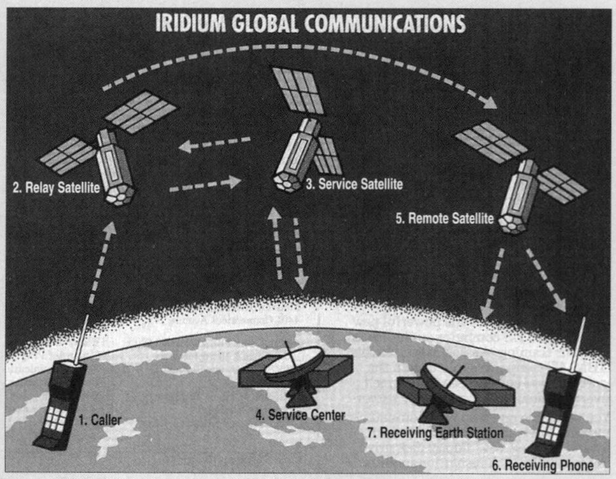

Motorola has developed a satellite-based system that allows callers in remote areas of the world to communicate by telephone. "You need to dial just one number to reach someone, even if you don't know where they are on earth," declared John Mitchell, vice chairman of the company, at a recent conference.

Mitchell says the system uses "a constellation of sixty-six satellites in low-earth orbit." Orginally the system was planned to have 77 satellites and is called Iridium, after the element Iridium, in which 77 electrons orbit the atomic nucleus. The satellites travel in polar orbits 413 nautical miles above earth, low enough to transmit signals to phones equipped with small antennas. Because of the large number of satellites, at least one is always in view.

Iridium is "really a cellular system turned upside-down," says Durell W. Hillis, corporate vice president and general manager for satellite communications at Motorola. "Each satellite projects a set of cells onto the earth. The earth-based users seem to be stationary, while the cells move rapidly through space."

In the Iridium system (see drawing), a caller

transmits both his phone number and serial number from a handset (1) to a "smart" satellite (2) capable of relaying the call to another satellite (3) orbiting over the caller's service center(4). After the centre checks the billing information and authorises the call, it beams a reply via the relay satellite to the original satellite, which sends the call to a satellite orbiting over the called party (5). The signal is beamed to that party's handset (6) or to an earth station (7) that sends the call through the local phone network.

By the time the Iridium system was ready for operation, most people in the United States and Canada had access to land-based cellular service. But the Iridium system, which provides service in remote areas, does not compete with cellular systems, according to Mitchell. Users include international business travellers, disaster relief workers, and passengers aboard planes or ships.

Motorola, which manufactured the satellites, handsets, and switching equipment for Iridium, put up 20 percent of the system's estimated $2.3 billion cost, Mitchell says.

Group B: Text 2

New Data Band Plays Nationwide

The company that first made nation-wide paging a reality with Skytel service is now working its wireless magic with PCs and electronic mail – with a little help from Microsoft.

Mobile Telecommunications Technologies, better known as Mtel, has built a two-way wireless data network serving the largest 300 urban areas in the United States. The company, under the subsidiary name Nationwide Wireless Network, covers the top 500 city areas. Microsoft has invested $30 million in the $150 million network and developed communications software for sending messages.

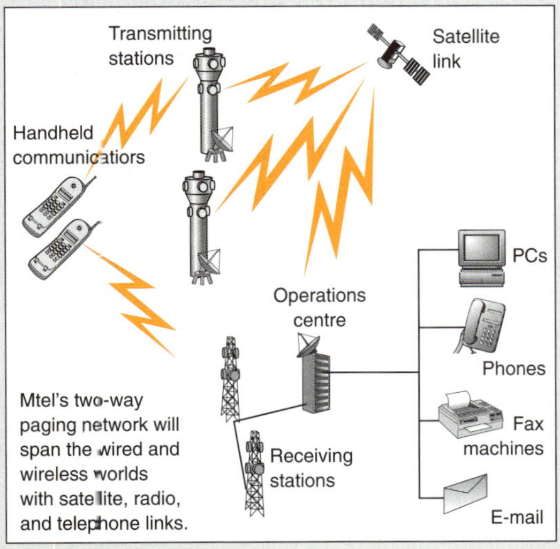

Mtel's two-way paging network will span the wired and wireless worlds with satellite, radio, and telephone links.

Unlike existing wireless data networks such as Ardis and RAM Mobile, the Mtel system is using separate transmitting and receiving stations to move messages between handheld communicators, laptop computers, and two-way data pagers. The network also links these devices to equipment on the wired phone network, such as fax machines, desktop PCs, and mainframe computers used by on-line services. The separate stations allow the network to use relatively few transmitting systems and many more receiving systems. With receiving systems always nearby, the portable devices themselves don't need as much transmitting power. Hence, the devices can be very small, and batteries should last 30 days and more.

The low-power approach also means that wireless communications devices don't require antennas. Mtel has worked with Motorola and others to build the communications equipment necessary for tapping into the network. In addition to two-way data pagers and custom communicators, Mtel says that PCMCIA cards can be used with existing portable computers.

FOCUS Features of technical texts

2 Noun phrases in technical writing

STUDY SKILL

In technical writing nouns are often used in groups (sometimes with other types of words) to express ideas succinctly. The most specific word is at the beginning of the group, the most general word is at the end. Noun phrases like this cannot usually be translated word for word.

Examples: a two-way wireless data network = a network for the transmission of data that uses wireless and works two ways (i.e. you can both transmit and receive)

a computer-controlled cabin noise-suppression system = a system to suppress noise in a(n aircraft) cabin which is controlled by a computer

- Now use the following definitions to "recreate" the noun phrases needed in the numbered gaps in the text below. (Ignore the underlined words in the text for now; they are not important till you get to section 3 below.)

1 a *connection* for *computers* used in *cellular systems*

2 a *cord* used for *telephones* which consists of *six wires*

3 *coaching* which was given over a *long distance*

4 the *number* for the *protocol* which makes *error control* possible

5 the *service* for *messages* received by a *computer* which operates in the *online* mode

6 the *systems* employed by *bulletin boards* which operate *electronically*

TELECOMMUNICATIONS

The … (1) sounds intriguing enough, but does it really work? To find out I ventured out of my ordinary office and into my minivan, where the makings of a four-wheel office were in place: an Audiovox car phone, a Toshiba portable computer, and a special cable that links the computer to the cellular phone.

I won't keep you on hold: it did work. Most of the time.

Because the optional Toshiba cellular kit wasn't yet in final packaged form and could support very few cellular phone models, I had to call three companies and also take a short trip to Radio Shack for some … (2) to gather all the necessary pieces. Making the proper connections without instructions also required some … (3). Presumably these inconveniences will be eliminated when the final product is available, which should be soon, according to the company.

Once wired up, I was onto the heart of the matter: Could I get on the cellular air-waves with my modem and stay there long enough to do something useful? I set up my communications software to dial the special … (4) for MCI Mail, an … (5). I figured I would need this feature to make a solid connection and maintain it. The reassuring sounds of the modem dialling were followed by the familiar screech of the connection. I was on. Better still, the connection held up well, enabling me to read and reply to a pair of electronic-mail messages, enduring only a

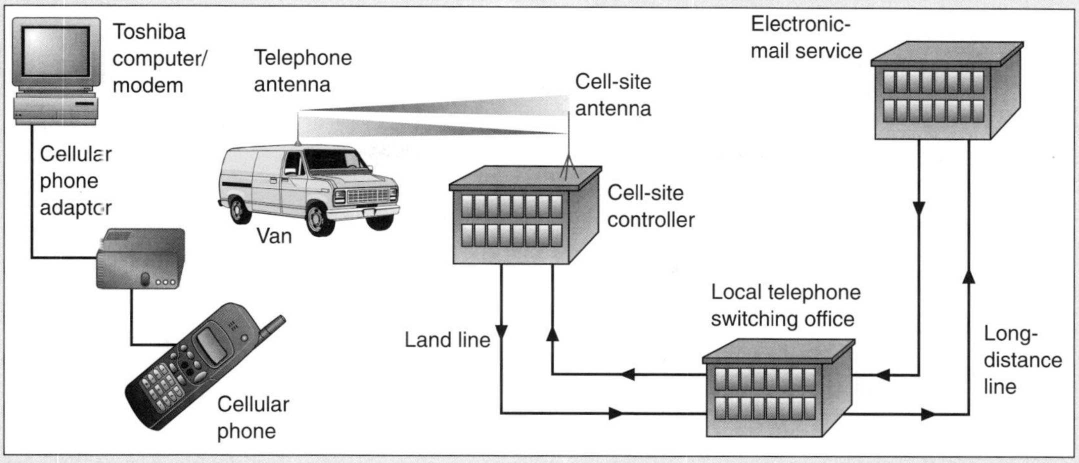

few "garbage" characters along the way. It was no fluke either; I successfully connected with MCI several times.

Next day I dialled the computer modem in my home office, which also supports the MNP protocol and was set to answer any incoming calls automatically. Again the results were consistently good. I was able to download, or receive, several files without a hitch.

Was this cellular data stuff going to be easy? No such luck. My calls to the CompuServe information service were repeatedly rejected. The next day I learned that only some of the service's entry numbers support the MNP protocol. The one I had been dialling did not but the correct number worked fine, though transmission was slower than usual on some calls.

Calling … (6) proved to be even harder. It appeared that none of the systems I attempted to contact used error-controlling modems. So it wasn't surprising that my success rate was only a little better – and not all the connections were noise-free.

All of this was in my driveway, mind you. On the road, there were good spots and not-so-good spots, which probably depended as much on the time of day and the number of people using their cellular phones as it did on the location. My location has level geography and good cellular coverage, something not every area can boast; people in areas with more hills or spotty cellular coverage may experience more glitches.

On the whole, the cellular-computer connection was surprisingly solid. With the right hardware and a little trial and error, it transformed a family vehicle into a van for all seasons. I'm not sure my wife is pleased with this development.

TELECOMMUNICATIONS

3 Formal and informal language in technical writing

STUDY SKILL

Technical texts which are written not for scientists or technical experts, but for the informed user of a particular technology (for example, computer magazines, magazines for hi-fi enthusiasts or hobby electricians) often mix formal language with rather informal language. This can cause difficulties in translations.

How would you translate the underlined parts in the text above?

Appendix

Pairwork tasks

Tapescripts

Vocabulary: chronological list

Vocabulary: alphabetical list

Mathematical operations

SI/ISO units, abbreviations and conversion table

Pairwork tasks

TASK 1

Unit 2, page 21

A You are Rajib Singh. You want to speak to Mr Swinson again because you need further information on an energy survey. You want to speak to him personally. You will be in your office for another 30 minutes, so you can call again. You dial his number.

TASK 2

Unit 2, page 22

A You are Rajib Singh. Your company specializes in importing and exporting computer accessories and software. You have had a lot of success and are now thinking about employing more people. The computer industry in general, however, is in a very difficult situation. Prices are falling and there is a lot of competition. Tell your friend Michael Bristow (= Partner B) about your situation and ask him for advice.

TASK 3

Unit 3, page 26

A You are Mr Toll. You have just come out of a meeting with your new boss. The diary below shows the appointments you have in the coming week. Now the phone rings. Answer the phone.

	MONDAY	TUESDAY	WEDNESDAY	THURSDAY	FRIDAY	SATURDAY
9 – 11	9 – 10 Dentist	9 – ? Board meeting	8.30 – 10.30 Oxford, Mrs Crewe	9.00 – 11.00 Meeting with Paul??	10.30 Phillip +	11.00 ↓
11 – 13				11.15 R+D briefing on BSOP	Les ≈ 12.00	↓
13 – 15		14.00 Meeting with Liza		Meeting with John?		↓
15 – 18			Call Playhouse for tickets (Friday/ Thursday)	17.30 Tennis Playhouse?	Playhouse?	16.00 Tennis

TASK 4

Unit 3, page 31

A Use the information below to prepare an explanation of the technical problem and how it was solved. Present your result to your partner.

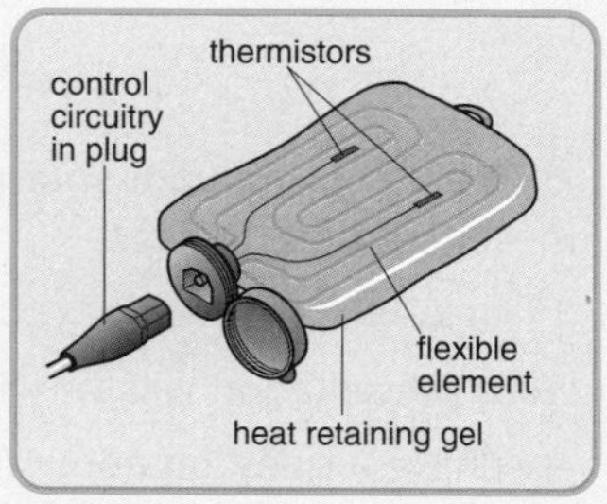

Safeheat, the market leader in hot-water bottles is introducing its second generation of "dry" hotwater bottles. Normal hot-water bottles can be difficult and dangerous to fill for anyone with arthritis or weak muscles. The Heat-Helper II looks like a normal hot-water bottle but works in a completely different way. A flexible heating element within the bottle heats up a heat-retaining gel, similar to that used in cooling blocks for picnic bags. The gel will not get as hot as water, but will keep its heat longer. While the old model was rather difficult to "set" to a certain temperature, improved controls in the plug allow exact setting of the desired temperature. Also, the new gel NPS12 is guaranteed to remain soft for four years. The tendency to harden with time, which the previous model experienced, will be a problem of the past.

TASK 5

Unit 3, page 33

A Describe the graph on the right showing manufacturing output in 1998/9. When you have finished, draw a graph of the down traffic for lifts in an office building according to your partner's description.

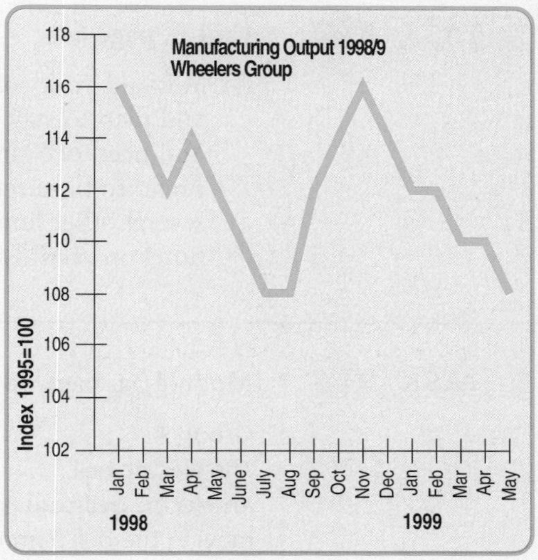

TASK 6

Unit 5, page 46

Situation 1

A You are at an exhibition. You are the representative for a company that has developed a new machine to identify and sort out different plastic materials. However, you have run out of brochures. There is just one left but you want to keep it until the new brochures have arrived.

Situation 2

A A colleague (Partner B) suddenly asks you to pick up a customer at the airport. Because you have promised to have lunch with a friend in town, you would prefer he asked somebody else.

TASK 7

Unit 5, page 50

A You are a member of the audience at the recycling presentation. Look back at sections 4 (Listening) and 5 (Reading), then choose phrases from the table below to ask the presenter about the recycling plant in Landshut. You can add other questions if you wish.

Sir/Madam, I thought your presentation was very interesting, but ...	Did you say that ...
I was very impressed by your presentation, but ...	... they recycled all the coolant and fluids needed during production?
Your presentation was very interesting, but ...	... the mechanical components such as engine and gearbox were also shredded?
... I have a question about (your figures).	... they burned the tyres and the rubber parts in Landshut?
... I didn't quite get whether you said (70%).	... they wouldn't use any plastic parts any more in the future
You said 9 million cars are taken off the road in Germany per year. That figure seems a bit high to me.	

TASK 8

Unit 6, page 61

A You and your colleague are going to attend the Motor Show in Birmingham and plan to stay for three nights (Mon. 12th April – Wednesday 15th April). You need two single rooms. The rooms should be reasonably priced. You would prefer to have rooms with showers and televisions. You have already phoned several other hotels, but they are all booked up. Now you are calling the last hotel on your list, the Bridgeport Hotel.

TASK 9

Module 3a, page 88

Group A

The electric bell

An electric bell contains an electromagnet that switches itself off and on very rapidly, moving the bell hammer as it does so. When the bell switch is pressed, current flows through the electromagnet and the hammer is pulled across to strike the gong. The movement pulls the contacts apart, which cuts power to the electromagnet – the hammer springs back, the contact closes again and the process repeats itself until the bell switch is released.

TASK 1

Unit 2, page 21

B You are Mr Bicknell. You are in Mr Swinson's office while Mr Swinson is in the warehouse. The phone rings and you answer the phone. It is Mr Singh.
You call the warehouse, but the line is engaged.
If Mr Singh calls again, you are still in Mr Swinson's office. You don't know where Mr Swinson is, but you expect him any minute. You offer to get Mr Swinson to call Mr Singh and you ask for his extension number.

TASK 2

Unit 2, page 22

B You are Michael Bristow. You have been in the computer industry for some time and Mr Singh (Partner A) is asking you for advice. Let him explain the situation. Suggest possible developments, using the phrases in the box on page 22. You can choose your advice from the following alternatives:

a Expansion in the same product range, specializing perhaps in computer games
b Specializing in small-scale software development, only limited and very specialized hardware
c Specializing in software sales and developing a consultancy for software application
d Other possibilities?

TASK 3

Unit 3, page 26

B You are Mr Grundland. The diary below shows the appointments you have in the coming week. You dial Mr Toll's number and wait for him to answer.

	MONDAG	TISDAG	ONSDAG	TORSDAG	FREDAG	LÖRDAG
9 – 11	*10.00 Meeting with Transcom*					
11 – 13	*(Prep. files)* ↓	*Meeting with Lars*	*12.00 Meeting with Public Relations*			
13 – 15	↓		↓		*14.30 Flight to Rome* ↓	
15 – 18	↓	*15.00 – 16.30 Lars + Jörge*			↓ *Hotel Siena*	

TASK 4

Unit 3, page 31

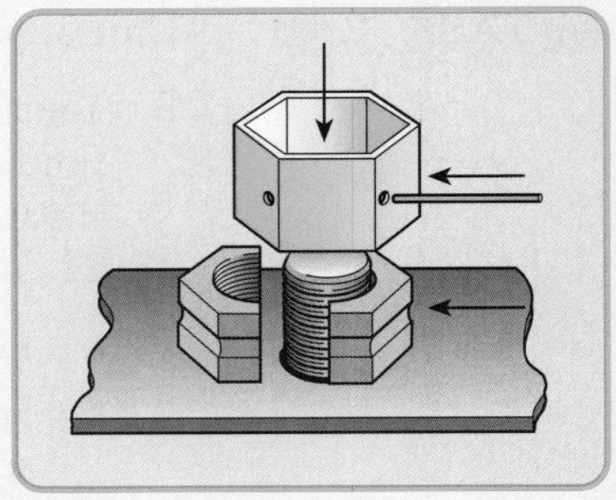

B Use the information below to prepare an explanation of the technical problem and how it was solved. Present your result to your partner.

A new sort of nut is being used in high-risk areas on Esspo's North Sea oil platforms. The company says that the new "sleeve nut" is the answer to problems in the past, when machinery could not be fastened and unfastened quickly enough. On several occasions the company lost valuable machinery in fires and heavy storms. The invention also saves time and money because even routine maintenance is easier in areas where corrosion affects ordinary steel nuts and bolts. The "sleeve nut" comprises two halves of a nut held together by the "sleeve", which is made of strong non-metallic material. To fasten the bolt, the two halves of the nut are placed around the bolt and the sleeve is positioned over them. Now the nut can be tightened. To unfasten the bolt, the sleeve is simply taken off the nut, and the two halves will come away from the bolt in a few seconds.

TASK 5

Unit 3, page 33

B First draw the curve representing manufacturing output in 1998/9 according to your partner's description. Then describe the down-traffic graph (see section 10, page 32) to your partner who will draw it according to your explanation.

TASK 6

Unit 5, page 46

Situation 1

B At an exhibition you are interested in a new machine that identifies plastic materials. There is only one brochure on the counter and you ask the representative (Partner A) if she/he can let you take it.

Situation 2

B You are supposed to pick up a customer at the airport at two o'clock . Now you realize you have a doctor's appointment at 2 o'clock. You ask a colleague (Partner A) to go to the airport for you.

TASK 7

Unit 5, page 50

B You are the presenter. Your partner will probably ask you about the following aspects of your presentation. Clarify misunderstandings using the phrases in the box on page 50. Read the presentation again, if you don't know the answers: for the first part of the presentation you can turn to page 137 to read the transcript.

- How many cars are taken off the road in Europe per year?
- What was done with the fluids needed during the production of a car?
- What was done with the main mechanical components of a car?
- What happened to the rubber parts of a vehicle?
- Would plastics still be used in the future?
- ...

TASK 8

Unit 6, page 61

B You are the receptionist at the Bridgeport Hotel in Birmingham. Because of the Motor Show, the hotel is almost fully booked from April 10th through to April 18th. There are one single and two double rooms left.

Double A	Double B	Single
two individual beds telephone no television shower/ no toilet price including breakfast: £61	en suite one kingsize bed television telephone price including breakfast £95	no shower / no toilet television telephone price including breakfast £43

TASK 9

Module 3a, page 88

Group B

The electric coil meter

If current passes through a coil in a magnetic field, the coil tries to turn. This is because the coil becomes a small electromagnet and tries to line up with the field in the same way that a compass needle does. If this turning effect is resisted by a small spring, the coil makes a simple meter. The greater the current through it, the more the coil turns against the spring. Most moving coil meters use a large number of coils to make them more sensitive. They also have extra features designed to make the scale numbers evenly spaced.

TASK 10

Module 4a, page 99

Complete the crossword below with your partner's help. Each of you has an incomplete version of it. There are no clues – you have to give your partner definitions of the words in your version (don't read the words out!), and complete the rest of the crossword from the definitions your partner gives you.

Puzzle B

Unit 6, section 7: sample solution

Dear Mr Worthing

Ref: ENA00105

I am writing in reply to your advertisement in The Times of October 19th and would like to apply for a position as a Technical Support Engineer.

I am a German national and have just graduated from the University of Applied Science in Dieburg with a diploma in applied electronics. During the last few years of my studies, I specialized in telecommunications, a field that I have always had an interest in and which I entered as an apprentice in telecommunication electronics. After my apprenticeship I spent four years working in the industry before I decided to go to university.
Having successfully completed the obligatory one-year preparatory course, I enrolled as a full-time student at the University of Applied Science in Aachen. In my third year I took the opportunity to continue my studies as an exchange student at King's College in Plymouth. On my return from England, I transferred to the University of Applied Science in Dieburg because it offered a more specialized course in telecommunications. Throughout my studies I have continued to work for my previous employer on a part-time basis. This also gave me the opportunity to carry out a research project with the assistance and close cooperation of the company.

I believe that my studies and my practical experience would provide a good basis for the position. I have always enjoyed working with people, and would welcome the challenge of learning as well as assuming responsibility.

Enclosed you will find my CV. I would very much appreciate the opportunity to discuss this application with you at any time.

Yours sincerely

Christian Rüteling

Tapescripts

Unit 1 – 1/7

First of all let me thank you again for having invited me and, of course, for your interest in our products. Our company tour has been very interesting for me and has helped me understand where you are and what your plans are. Before I come to a more detailed presentation of the products you expressed an interest in, I think it would help to outline briefly some more general concepts of manufacturing systems.

Now, from what I have seen and from what you have explained to me, I understand that the majority of parts in this plant are manufactured in what is commonly called a job shop. In job-shop systems we have groups of more or less identical machinery located in different areas of the plant so that the parts to be machined go from one station to the next. You see a typical layout of a job-shop system in Figure 1. As you can see, different parts move through the plant in different paths. Part A, for example, first goes to the turning lathes, is then taken to the grinding machines before it reaches assembly and storage. Oh, I'm sorry, I've just realized that I've forgotten to mark the lathes. The lathes are the boxes between reception and the grinding department. Now, let's look at part C. Part C first goes to the saws and then to the drill presses. From the drill presses it is brought to the grinding machines, before it is heat treated, assembled and stored. The diagram shows clearly that there is a lot of transport time involved, as of course you know.

Let's now turn to Figure 2, the flow-job system – or assembly line –, the kind of system you use for the manufacture of door parts. Such a system is ideal for the mass production of one and the same product. The diagram shows very clear production lines. The parts "flow" from each station to the next, and this of course reduces transport time drastically. The disadvantage of such a system is, as you know, that it takes a lot of time to set up the production line for a new product. And this is where you are facing problems now. The trend in modern production clearly goes away from mass-produced identical parts – to mass-produced dissimilar parts, even in the car industry.

This brings me to Figure 3, the diagram of a cellular manufacturing system. The cellular manufacturing system groups different kinds of machinery together which are needed to complete a product. As you can see from the diagram, the machinery is grouped according to the sequence of operations needed to complete a product. It is thus not unlike an assembly line, but much more flexible. The individual machines are typically NC or CNC machines, which means that production can be changed over from one product to another – almost immediately. The cell shown in Figure 3 is U-shaped so that the operator can easily control all the machinery.

Finally, I'd like to draw your attention to Figure 4, the example of a robotic cell in which manufacture is completely automatic. In such a cell a robot takes the place of the operator, which, of course, puts very high demands on the capabilities of the robot. This is ideal for working environments that are not suitable for people to work in, or if the parts to be moved from one machine to the next are very heavy. In your company both conditions come together in the casting department.

I think it's fair to say that cellular manufacturing systems are the systems of the future. Apart from the high degree of flexibility, accuracy and manufacturing speed which they offer, they are also necessary elements of modern management systems. We will be talking about this again a little later. Now, I think this brief overview of different types of manufacturing systems helps us to see where ABA's products nicely fit into your plans. But before I come to that, are there any questions or comments?

Unit 2 – 2/8

Mr Swinson: Good morning, Mr Singh. How are you?
Mr Singh: Fine, thank you. How about yourself?
Mr Swinson: Very well, thank you. Sorry that I wasn't able to come round today. We're completely snowed under this week.
Mr Singh: That's O.K. Now, tell me what you've come up with.
Mr Swinson: Well, as you already suspected, you'll need a new boiler. But that's not the only thing. There are a few other things you might want to consider.
Mr Singh: Oh, what are they then?
Mr Swinson: Well, let me start with the boiler. You could simply replace the existing oil-fired one with a newer model, but I suggest you exchange it for a gas-fired one. They're not very much more expensive.
Mr Singh: How much would that be?
Mr Swinson: Gas-fired condensing boilers are approximately £500 more expensive than traditional oil-fired ones.
Mr Singh: Hmm.
Mr Swinson: But the boiler will pay for itself in five to six years. You can expect an annual saving of about £60 to £80.
Mr Singh: Oh, that sounds O.K.
Mr Swinson: Now, the piping and the radiators are in surprisingly good shape.
Mr Singh: Well that's good to hear.
Mr Swinson: But the pipes aren't properly insulated. And the radiators aren't equipped with thermostats.
Mr Singh: Well, I knew about the thermostats.
Mr Swinson: I would certainly recommend insulating the pipes with 2 inch thermofill. That's not very expensive. That'll probably cost you between £100 and £150.
Mr Singh: How about the thermostats?
Mr Swinson: Modern thermostatic radiator valves are standard in new buildings. They are about £10 a piece.
Mr Singh: Hm, is there anything else?
Mr Swinson: I'm afraid so. You see, the house is 27 years old, and the windows and the walls don't conform to the new building regulations any more.
Mr Singh: Is that right?
Mr Swinson: Yes, but that's only for new buildings. In your house you could leave everything as it is. The thing is, however, you could save almost 15% of your annual heating costs if you installed new windows. And if the walls and the loft are properly insulated, that'll cut your costs by another 15 per cent.
Mr Singh: What about the initial costs?
Mr Swinson: That's difficult to say. I suspect we're talking about £1200 for double-glazed windows and £1500 for the wall and loft insulation.
Mr Singh: Hm. That seems rather a lot.
Mr Swinson: Yes. But the savings are considerable. I'd give it a thought. But I don't want to go into details now. I'll send you my report and some general information. Why don't you have a look at it, and we'll get together next week and talk about it.
Mr Singh: O.K. Well, thank you very much, and I'll see you next week then.
Mr Swinson: Yes, goodbye, Mr Singh.

Unit 3 – 3/1

Mr Bretton: Hello, Bretton speaking.
Mr Olafson: Good morning Mr Bretton, this is Jan Olafson from Hansen in Sweden. Thank you very much for sending the brochures. They are very interesting.
Mr Bretton: Are they? I'm glad to hear that – and I apologize again for not sending the material on drives, but I promise to send it as soon as possible.
Mr Olafson: No hurry. Actually, I am planning to visit the UK in June, and this is really why I'm calling. I was wondering whether we could find the time to meet while I'm in the area.
Mr Bretton: Well, that's good to hear. June you said ... there should be no problem, I'll be in Manchester the whole month. What time are you thinking about?
Mr Olafson: It's really the first week in June that I'll be in the Manchester area.
Mr Bretton: So, what about Tuesday or Wednesday morning then?
Mr Olafson: Hm. I'm afraid that won't be possible, I've already arranged a meeting in Sheffield, but I'll be free Tuesday afternoon.
Mr Bretton: Unfortunately, I won't be available then, I have a meeting here in the office. But how about early Thursday morning?
Mr Olafson: No, sorry, I have another meeting on Thursday morning but I'm free again in the afternoon. How about that, Thursday afternoon?
Mr Bretton: That sounds alright. I've got nothing planned for Thursday afternoon. Oh, wait a minute, I've just realized that I have a tennis appointment at 6 o'clock. What time could you be here?
Mr Olafson: Fairly early, let's say 2 o'clock? Would 2 o'clock suit you?
Mr Bretton: 2 o'clock sounds fine with me.
Mr Olafson: So it's two o'clock then.
Mr Bretton: Will you be coming by car?
Mr Olafson: Yes, I will.
Mr Bretton: I'll send you directions – your fax number is still the same, isn't it?
Mr Olafson: Yes of course, it's still the same number.
Mr Bretton: I'll send it straight through to you. And I'm looking forward to seeing you.
Mr Olafson: So am I, goodbye Mr Bretton.

3/6

Mr Olafson: ... and as you know, we're bidding for the lift contract for the new European Microsound office. The one to be built in Malmö.
Mr Bretton: Yes, I remember you mentioning it. Quite a big contract, I imagine.
Mr Olafson: Yes, it is. And quite a challenge for us. Winning it could easily be a breakthrough in the Scandinavian market. But the plans were changed to a three-lift system instead of the four that we had been working on.
Mr Bretton: Hmm. I can see that that is a major headache. What does it mean for you?
Mr Olafson: We were doing pretty well up to that point. You know that one of the contract requirements was to come up with a design which would have at least 15% lower running costs.
Mr Bretton: Yes, and you mentioned problems with your motor supplier.
Mr Olafson: Yes, but we thought we'd solved them. With the new design we are in real trouble.
Mr Bretton: Why's that?
Mr Olafson: Well, you see, initial costs will of course be lower. Not so much as one would think, but still. But for the running costs we'll have to start again.
Mr Bretton: Hmm.
Mr Olafson: First of all, the individual cars will have to be bigger and thus heavier. Acceleration time, however, will have to be higher if we don't want to increase the response time to a call too much. This, of course, will increase energy consumption for each individual car. What it means for the whole system is a different question. But the real snag is that it places completely new demands on the lift motors.
Mr Bretton: Yes, I can see that.
Mr Olafson: Not only in terms of power consumption, but also in terms of control, emergency regulations, acceleration capabilities and ... well that's basically it.
Mr Bretton: But that was already a problem with the original design, wasn't it?
Mr Olafson: Yes, but one we had solved. We'd got around the problem of fire regulations as well.
Mr Bretton: Hmm.
Mr Olafson: Now as far as the engines are concerned, what I'd like to discuss with you is this ...

Unit 4 – 4/5

Secretary: Europäische Architektenkammer, guten Tag.
Mr Watts: Hello, dies is Mr Watts aus England. Kann ich bitte mit Herr Schlueli sprechen?
Secretary: Ja, einen Moment bitte, ich verbinde.
Mr Schlueli: Hallo Mr Watts. How are you?
Mr Watts: Ah, good – English. Mein Deutsch ist sehr schlecht.
Mr Schlueli: But I think your German is very good.
Mr Watts: No, not really. Just the necessary standard telephone phrases. I used to speak it much better, but it's rather rusty now.
Mr Schlueli: Hm, I know what you mean. I lived and studied in Italy for a while when I was a student, but that was a long time ago. I have to go there next month. It will probably be difficult.
Mr Watts: Yes, it's difficult to stay in practice, isn't it?
Mr Schlueli: Yes, it is. I suppose you're calling about your visit to the Zumikon building.
Mr Watts: Yes, that's right. We're very grateful to you for having made a visit possible. I hope we won't be taking too much of your time.
Mr Schlueli: No, not at all, it's a pleasure. I'm looking forward to showing you the building with the architect. He was very much involved in its design and enjoyed it very much. I think he can tell you quite a bit about it. Actually you'll be talking to the owner as well. He's a very interesting person and very proud of the building. So he'll be very pleased to show you around.
Mr Watts: I'm very glad to hear that. Now, what day would be most convenient? We're very easy.
Mr Schlueli: The owner said Tuesday would suit him best. And that's fine with the architect, and me, too.
Mr Watts: So, Tuesday, Tuesday the 13th. Yes that's actually very good. And at what time?
Mr Schlueli: Would the morning or the afternoon be better for you?
Mr Watts: Whichever you prefer. We'll probably be arriving on Monday evening, so any time on Tuesday would be OK for us.
Mr Schlueli: Well, let's say 10 o'clock then.
Mr Watts: Good, ten o'clock. Now could you recommend a hotel? I haven't been to Zurich for quite a while, and ...
Mr Schlueli: We usually recommend the Hotel Geneva. It's not very big, but very comfortable, reasonably priced and the restaurant is excellent.
Mr Watts: That sounds good. You don't by any chance have the telephone number, do you?

Mr Schlueli: Not right now, but we'll fax it to you. We can also make a reservation for you, if you like.
Mr Watts: No thanks, that won't be necessary.
Mr Schlueli: Ok. So it's Tuesday, the 13th, at ten o'clock then.
Mr Watts: Yes, and thank you very much again.
Mr Schlueli: My pleasure. Goodbye, Mr Watts.
Mr Watts: Goodbye.

4/7

Mr Schlueli: OK, now here are the plans. As you saw, the house is integrated nicely into the hillside.
Mr Watts: Yes, and I particularly liked the series of gardens and terraces. It gives the house a very open and friendly atmosphere. Almost like ...
Mr Fielding: ... like a village, yes it has a village-like character.
Mr Watts: Yes it does, doesn't it?
Mr Schlueli: This was actually not the initial plan but was partially forced upon the architects. The building code in the area sets very strict limits to height and mass. But when they realized that they were talking about more than 10,000 feet of floor space they knew that they couldn't just build a house. They had to have a fundamentally different approach.
Mr Watts: Yes, I see. And basically the house is grouped into two main wings ...
Mr Schlueli: ... which are connected by the dining room. Now if we take this terrace as our starting point – this is actually a sculpture terrace where they normally display sculptures. And there's a swimming pool that blends in very nicely, too.
Mr Fielding: That's this to the left here?
Mr Schlueli: That's right. Now if we go inside and down to the first level, this is the art gallery.
Mr Watts: Ah yes.
Mr Schlueli: The last time I saw the house there were paintings by Luciano Silvestri, a Swiss artist, and an original Miro in the semicircular extension. That part, the semicircle, becomes a full circle on the second level where the library is.
Mr Fielding: Ah yes, I see.
Mr Schlueli: And above this main part of the gallery is the family dining room.
Mr Fielding: That must be a very spacious dining room then.
Mr Schlueli: Yes, it is. But the art gallery receives its only natural light from above, through the glass ceiling, so not the whole ceiling of the art gallery is the floor of the dining room.
Mr Watts: I see.
Mr Schlueli: The other rooms on the first level are mainly storage rooms and a wine cellar, and of course a bathroom.
Yes, and there's a conference room, too.
Mr Fielding: Right.
Mr Schlueli: OK, now here on the second level is the library, you see that it's almost circular. Actually, this is the reading room part of the library. The shelves are in this part over here.
Mr Watts: I see.
Mr Schlueli: Actually it's one of the most inviting libraries I have ever seen, cosy and at the same time, what is the word, academic.
Mr Fielding: It would have been nice to see it.
Mr Schlueli: Now, before, as it were, we go to the living room, let's go along here through the dining room to the kitchen wing.
Mr Watts: The dining room is big, isn't it?
Mr Schlueli: Yes, very spacious. It's a large family and they often have guests it seems, so ..., and this is what I meant with the dining room giving light to the art gallery below.
Mr Watts: Yes, I see. There are windows on both sides.
Mr Schlueli: Yes one side opens to the terrace, with a view to the Alps in the distance, and the other gives you a splendid view of Zurich. Right, and beyond it, at the northern, north-eastern end, is the kitchen with a breakfast area.
This wing has a guest room as well. The kitchen is very inviting, chosen with a lot of care. It seems they spend a lot of time there.
Mr Fielding: Hm.
Mr Schlueli: Now if we go back through the dining room, as it were, the living room opens to the left at the end.
Mr Watts: Another very large room.
Mr Schlueli: Yes, it's hugely impressive, with a factory-style design open fireplace, robust and at the same time very elegant. Now at the back of the living room there are two bedrooms with an open space in between. And a bathroom.
Mr Fielding: And more bedrooms on the third level, I suppose.
Mr Schlueli: Yes. There are three more bedrooms on the third level, and the master bedroom ...

Unit 5 – 5/1 – *Conversation 1*

John: Hi Brian, how's your project coming along?
Brian: Glad you ask, I meant to talk to you about it. Look, I don't think I will be able to get the results in time. Things look promising, but I suspect I will need a little more time.
John: Hmm. How long do you think you'll need?
Brian: Say a week to ten days more.
John: I'll talk to Robert about it. But there is something else I wanted to talk to you about. Have you seen the congress announcement in *Engineering*?
Brian: Which one?
John: The Autotech in Birmingham, in November.
Brian: No, I didn't notice it.
John: You see, they're having a series of speakers on the recycling issue. That should be very interesting, shouldn't it?
Brian: Are they really? Yes, that should be very interesting.
John: Yes. Now, it's still a couple of months ahead of us, so what I meant to ask you is if you'd like to go there. It could be helpful to us, also in making contacts. You have to go to Manchester anyway, don't you, for the ...
Brian: Yes, but that's in September, which isn't really that close to November. But I guess I could change that. What dates are you talking about?
John: The 12th to the 15th.
Brian: Let me see. ... Ah, sorry John, that's exactly the time when my wife won't be home. She's attending a conference herself and someone has to be with the kids.
John: And you don't think you could talk to her about that?
Brian: I guess I could, but I think I already know the answer. You see, she's hoping for a promotion and, well you know how that is.
John: And you don't think you could arrange something? I mean it could prove very important for us to have someone attending the conference and ...
Brian: No, really John, I don't think there is anything I can do about it. Couldn't you ask Lisa?
John: Well, you know Lisa! She hates conferences.
Brian: Yes, I know, but try anyway. I really am sorry, John, but I don't think I can make it.
John: OK, let's drop it then. There is still plenty of time so we can talk about it later. Oh, by the way, could you send me a preliminary report on your project, so that I can discuss the delay with Robert?

Conversation 2

John: Hi Lisa, this is John.
Lisa: Hi John, how are you? Have you had a nice vacation?
John: Oh yes, excellent, how about yourself?

Lisa: I haven't been on vacation yet. Remember, Jack took his family to Canada and someone had to be here.

John: Of course, sorry about that. Yes, I'm very glad you stayed. Hmm. I've just talked to Brian. You see, there's this congress on recycling in the automotive industry in Europe, and we felt it would be very good if one of us went there. Now, Brian suggested you'd be the best person to go. He says that there is probably quite a bit on the plastic issue and that's really your field. Also, some presentations will be in French, and from what I hear, your French is excellent.

Lisa: A bit rusty right now, but you know that ...

John: Yes, I know that you've got plenty of work to do. But it's still some time ahead of us and three days won't hurt, will they?

Lisa: No, but you know that I don't like going to conferences.

John: Oh, yes, I forgot. But I'm sure you'll find it very interesting and it could really help us a lot in making contacts ...

Lisa: Which is precisely why I don't like conferences. I hate this constant small talk.

John: But you're very good at it. I've had all kinds of people telling me how interesting they found it talking to you. And anyway, it's only for three days. I'll send you the specifics this afternoon and we'll talk about it later. Oh, I almost forgot, could you let me have a word with Jack please? I need to ...

5/4

I think we all agree that the present situation with recycling in the motor industry can be said to be both positive and negative. It is positive in so far as about 75 percent of the weight of a car is already efficiently recycled, and this is much more than for most other products. Think of televisions, for example, or, even more problematic in recent years, the recycling of computers. However, the 75% represent more or less the metallic content of a car, and people are becoming increasingly worried about the remaining 25 percent. These worries are justified and understandable because 25 percent of the 9 million cars taken off the road in Europe alone means an enormous 500,000 tons of "light waste" alone. And the traditional way of dealing with this light waste has been to dump it in landfills. And, of course, the term "light waste" is very misleading, as the environmental problems that this waste already causes and will cause in the future are certainly not "light".

But positive developments are under way and have been for some time. Let's go back a few years first, to the time when BMW, as one of the first European car makers to take a fresh look at recycling, opened their pilot recycling plant in Landshut, Germany, to determine just what it would take to dismantle vehicles and recover their materials in large amounts. At the plant 1500 to 2000 cars were dismantled annually. The operation revealed a lot about the dismantling demands and the kind of infrastructure needed to transport different materials to the manufacturers or recyclers. But what was probably even more important was that questions were raised about the changes designers are now having to make to the way cars are constructed in the first place.

The results of the pilot study were documented in an environmental data bank which also recorded the environmental characteristics of all materials used during production. This meant that all the environmental characteristics of all the fluids such as coolants and lubricants needed during manufacture were recorded along with the materials' engineering characteristics. Data like this now makes it possible for the engineer who designs a new dashboard, for example, to get information about the environmental impact of the production process as well as what happens once the car is scrapped. Is it a problem as far as recycling is concerned or isn't it? Can he use this material, and is the company able to recycle it fully at the end of the car's lifetime?

Unit 6 – 6/4

Part 1

Mick: Hello?

Carol: Hello Mick, this is Carol ...

Mick: Oh hello, good to hear you. I haven't heard from you in a lifetime. How are things?

Carol: Fine, thanks, I'm doing all right, how about yourself?

Mick: Not too bad. I'm learning to change nappies. Something new and far more interesting than a lot of other things I've been doing recently.

Carol: Oh, right, the baby. So that's why nobody ever answers when I call.

Mick: Have you tried often?

Carol: Several times. This is the third time today.

Mick: So that was you about an hour ago. I heard the phone but I was busy tapping something into the computer while Linda was asleep and didn't bother to pick up the phone. Had I known it was you ...

Carol: So, it's good being a parent?

Mick: Great, I can recommend it. How about yourself then? How are your studies going?

Carol: OK. I've finally finished. I had my last exam three months ago.

Mick: I'm glad to hear that. You weren't doing too well the last time we talked.

Carol: No, I wasn't. I was struggling to get my dissertation finished, and a tough job that was. But it's over now. I've finally graduated.

Mick: Congratulations! What are you planning to do now? A trip to England, I hope, and coming to visit us.

Carol: Well, not quite, but almost. You see, I saw this advertisement in last Saturday's *Times*. There is a job with the British subsidiary of Telecom 3000 and I'm wondering whether I should apply. It says that they are hoping for applications from European countries.

Mick: Well that sounds good. What kind of job is it?

Carol: That's partly why I'm calling you. There are a few things I'd like to check before I apply.

Part 2

Mick: Yes, what are they?

Carol: Well, they are looking for Technical Support Engineers, Customer Support Engineers and Systems Engineers. You know our job descriptions are a bit different, so what do you think they mean?

Mick: Hmm, I'm not quite sure, but I suppose the Systems Engineers will look at the telecommunication systems as a whole. They are in charge of developing the system and will have no direct contact with the client. The Technical Support Engineer looks after the existing system, subsystems and components. He will have occasional contact with clients. And I think that the Customer Support Engineer is the one who deals with the public. He's the one who responds to problems experienced by the clients.

Carol: Yes, that's what I thought it would be. But I've got another question. An HND is a Higher National Diploma, isn't it?

Mick: Yes, it is.

Carol: And that corresponds to our university degree?

Mick: That's right.

Carol: Also, do you happen to know what the NT DMS systems are?

Mick: You know that telecommunications isn't really my field. I'm more in computing. I could guess. DMS could be something like Digital Monitoring System, but I'm really not sure. I suppose I could get the information and call you or fax it to you.

Carol: Would you? Thanks a lot, that would be a big help.

Mick: Anything else?

Carol: No, not really. I think that's it. I'll think about it. Look you won't be going anywhere anytime soon, will you?

Mick: Actually, we were planning to spend some time in Wales. You know my parents have got a cottage there. But we had to cancel it. It's really very busy in the office, so ...
Carol: Sorry to hear that.
Mick: Well, no problem, really. We'll do it in the autumn instead. Anyway, if there's anything else you need to know, just give me a ring.
Carol: I will. In any case, I'll let you know what happens. And give my love to Lisa.
Mick: I certainly will.
Carol: Bye, Mick.
Mick: Bye, Carol.

6/8

Phone call 1

Receptionist: Heartwood Hotel, how can I help you?
Carol: Yes, hello, hmm, this is Carol Mistadou speaking. Er ...
I'm calling to see if you have any vacancies from the 22nd of April to the 26th.
Receptionist: Let me see, the 22nd, you said.
Carol: Yes, the 22nd through to the 26th.
Receptionist: Would that be a single room or a double?
Carol: A single, please.
Receptionist: Hmm, I'm afraid I can't help you there. It looks as if we're booked up. Yes, sorry, we are.
Carol: And what about a double?
Receptionist: A double. Yes, I can help you with a double.
Carol: And how much would that be?
Receptionist: That's £75, including breakfast.
Carol: Hmm, that's a bit much. I think I'd rather try to find a single room somewhere else. But thank you very much.
Receptionist: You're welcome.
Carol: Goodbye.
Receptionist: Goodbye.

Phone call 2

Receptionist: Hello, Green Leaves Hotel, how can I help you?
Carol: Yes, I was wondering whether you have any rooms left in the fourth week of April.
Receptionist: Hmm. The weekend is booked, but we've got lots of rooms during the week. What dates are we talking about?
Carol: That's the 22nd through to the 26th.
Receptionist: That's Monday through to Saturday. You'd be leaving on Saturday?
Carol: Yes.
Receptionist: No problem then. Would you like a single or double?
Carol: Single, please.
Receptionist: Now, we have singles for £45, £57 and £68 a night, breakfast included.
Carol: And what's the difference?
Receptionist: The one for £68 is en suite with a bath. There's also a TV, a telephone and a minibar, which neither of the other two has.
Carol: Well, I wouldn't mind. What about showers?
Receptionist: Oh, all our rooms have showers. The one for £45 is nice, just under the roof with a good view of the cathedral. The one for £57 is more spacious. It's on the first floor.
Carol: Is the one for £45 reasonably quiet?
Receptionist: Yes, this is a very quiet neighbourhood.
Carol: I think I'd like to book the one for £45 then. Do you accept credit cards?
Receptionist: Yes, we do. So that's for the 22nd to the 26th. And the name was ...?
Carol: Carol Mistadou. That's M-I-S-T-A-D-O-U.
Receptionist: Could you confirm your reservation in writing, please?
Carol: Yes, of course. I'll send you a fax.
Receptionist: Fine. See you on the 22nd, then.
Carol: Yes. Thank you very much and goodbye.
Receptionist: Thank you for calling. Goodbye.

Module 1a – M1a/9

A very early example of an automated control mechanism is the automatic pressure vessel which was invented centuries ago. This device was either used to provide steam with a specific pressure or it served as a safety device. Both purposes were important in steam engines, for example. Now, as you can see, a lever is attached to the lid of the pot and the weight of this lever holds a plug in the hole of the lid. When the steam pressure goes above a certain level, the pressure against the plug is sufficient to lift it off its seat. Now steam is released and the pressure in the vessel can fall. Sometimes the lever was fitted with a weight which could be moved along the lever so that the pressure in the vessel could be adjusted.
Now, Illustration 2, the transistor, shows a much more modern control or servo-device which is used in electronics to control or amplify electronic signals. Transistors are, as some of you will know, semiconductors, and it is this semiconducting property of the transistor which allows a large current to pass through the transistor at one time and to totally block the passage of electrons at another time. In Illustration 2 a large current can be made to pass from the bottom to the top of the transistor, i.e. from the emitter to the collector. The thing is, however, that this flow of electrons from emitter to collector is only possible when a much smaller current passes from the emitter via the switch and a resistor to the positive pole. Obviously, the switch must be closed, otherwise no current will flow at all. In other words then, a small current from the emitter to the positive pole via a resistor allows a much larger current to flow from emitter to collector to the positive pole. Power amplifications of more than a hundred times are possible with this small, cheap and reliable device.
Now let's look at Illustration 3 now, which shows an automatic clamping device. The arrangement is used to automatically hold a workpiece in place whatever its shape, size, dimension or rigidity. With little exaggeration you could say that the workpiece could be an egg that we want to drill a hole into, which of course breaks easily, or something made from steel, which is much, much stronger. Now, what happens is this: when the workpiece is placed on the worktable, the hydraulic clamp moves down until it touches the workpiece and begins to put pressure on the workpiece. As you can see, a strain gauge is attached to the clamp column and this strain gauge measures the force that is applied by the column. The strain gauge now sends a weak signal to an amplifier, where the signal is amplified. The new signal is then converted to a digital signal in the Analogue-Digital Converter, or ADC for short. This conversion is necessary because from the ADC the now digital signal travels to the microcomputer, where the signal is interpreted and compared with a pre-programmed value. If the two values are different, the microcomputer sends a signal to the hydraulic valve, which increases or decreases the pressure in the hydraulic cylinder. With our egg, for example, the pre-programmed value would be very small, and the microcomputer would decrease the pressure early enough for the egg not to break, so that we don't get scrambled eggs.

Module 2a – M2a/5

The amount of useful heat produced by a solar hot-water system depends on two factors. Firstly, the input of solar heat, secondly the total heat losses of the collector, the heat exchanger and the connecting pipes. Let's talk about the input side first, and I'd like you to look at Figure A now. As you can see, the maximum possible energy which could be received by a collector at noon is a little more than 3 megajoules per hour. That translates into something like 900 watts. This is, of course, per square metre of solar collector.

For that much input we need a tracking-type collector, i. e. one that always points to the sun, and we also need a very clear sky. The total heat gain for such a collector is an impressive 12 kilowatts. Sadly enough, however, very clear skies are not known to be common in the north of France – nor anywhere else in northwest Europe for that matter – so curve B is a more realistic one. This is for an average summer day, and you can see that we've already lost close to 150 watts per square metre at noon. Because very clear skies are not typical for our area, rotating or tracking systems are not really economical, given their high initial cost. This brings us to curve C which shows the distribution of heat received by a flat fixed collector – the type you may see around here. As curve C shows, the collector will receive approximately 650 watts per square metre at noon, and the total energy received per day will only be about half of what a tracking system would get in ideal weather. Now, the next problem is that not all of the energy received by the collector will actually heat the absorber: some of it will be reflected away or absorbed by the glass so that we end up with curve D. The curve shows that the energy input rises to a maximum of just over 600 watts at midday, and the total heat energy received per day is around 5 kilowatt hours. That's already less than half of what a tracking system gets in ideal weather. But the really sad truth of Figure A is its mid-winter version, labelled D1. You will notice that the maximum solar energy received at noon on an average mid-winter day is something like 200 watts and the total daily delivery will be 1.25 kilowatt hours per square metre. Notice too that the system gains heat only from 8 a.m. to 4 p.m., as opposed to the 16 hours in the summer when it already starts collecting heat at 4 a.m.

Module 3a – M3a/6

The graph in Figure 4 illustrates the alternating current produced by a simple a.c. generator. It shows how the generation of current during a complete rotation of the coils varies according to the coil position. With the coil starting in a horizontal position relative to the poles of the magnet, no current is generated. This is because in this position the magnetic lines of force are being cut at maximum speed. As the coil continues its rotation in a counter-clockwise direction, generation of current builds up gradually until maximum current is produced a quarter of a revolution later. During these first 90 degrees the current is taken as positive, but now the direction reverses and for the next 180 degrees there is a steady decrease of current generated until zero current is reached before the direction is reversed again for the last 90 degrees. The graph in Figure 4 shows a perfect sinewave and positive and negative peaks of the curve are referred to as maximum forward current and maximum reverse current.

Module 5a – M5a/3

In this pilot plant we process approximately 10 tons of recyclables per day. The recyclables, they are glass, tin and aluminium cans, as well as all sorts of plastic containers, are collected in the community once a week in special 12 gallon containers. People are pretty good about what they put into the containers, although we do occasionally get paper and plants and things like that. We've even had dead animals. But that can't really be avoided. Anyway, they are dumped on a conveyor belt and are first checked manually for materials that absolutely don't belong there and are then tipped onto an inclined electromagnetic conveyor belt. As the materials travel upwards, ferrous materials attach themselves to the belt while others, like aluminium, plastics and glass slide down into a hopper which directs them to another belt. As they travel along the belt, the glass is sorted out. This is done manually and the glass is separated by colour. Next comes the aluminium. To sort out the aluminium, we pass it through a kind of gate where an eddy current magnet induces an alternating charge in the aluminium articles which makes them hop off the conveyor belt. Don't ask me how this works, I'm not an expert in this, I don't really know myself. So, now that we've got rid of the glass and aluminium, and of course the ferrous material, we are left with the plastics. For the plastic materials we have a fairly sophisticated sorting system although the first step in plastic sorting is rather old-fashioned because we manually sort out certain plastics which are easily identifiable. Easily identifiable plastics are, for example, beverage bottles made from PET and some other containers such as yoghurt containers. The rest is shredded into strips, then washed and finally ground to half inch flakes. The flakes then proceed to the hydrocyclone. A hydrocyclone is a large vertically oriented cylinder in which pumped-in water spins around the side, just like in a cyclone. We add special additives to the water which changes its density so that the lighter flakes spin upwards and exit the cylinders together with the water while heavier flakes drop out. The plastic flakes are now dried and taken to a machine that turns them into small pellets before they are shot into a shipping box for sale to other companies which process them further.

Module 6a – M6a/6

Travellers in the first and business class of several international air carriers can already now enjoy the comfort of listening to music in the quality they are used to from their hi-fidelity equipment in their homes. Their headsets are equipped with electronic earphones rather than the traditional air-based earphones, allowing not only far better sound transmission but also the choice of more channels. Some carriers have now also begun to install personal video monitors. This improves the quality of the video image and will eventually provide a choice from a video library. And for the more work-driven high-flier, seat-based power sockets will soon mark the end of the times when laptops lost their blinks as batteries went dead.

Not everyone aboard a plane, however, will cherish the latest improvements in entertainment or laptop facilities. Some just want to take a refreshing nap or have a friendly chat with the person next to them. Particularly travellers on commuter planes driven by turboprop engines have found that noise from the engines proved a big obstacle to their comfort. This may soon be a problem of the past if a new invention by the European aircraft manufacturer SAAB holds up to expectations. According to a SAAB spokesman, the company has begun installing computer-controlled cabin noise-suppression systems in some of its commuter planes operating on the continent. Microphones throughout the cabin pick up various sounds, and a computer combines these with engine speed information to produce sound waves that are exactly opposite to the undesirable noise. These sound waves, technically 180º out of phase with the original noise, are projected into the cabin through speakers, so that the valleys and peaks in the original noise waves are cancelled by corresponding peaks in the suppression sound. A controller fine-tunes this zero signal. According to SAAB, the system has already been installed in several of its aircraft flying in Europe and is expected to be installed in some planes operating in the US. Other manufacturers, he said, will also have similar setups in the years to come.

Vocabulary: chronological list

Unit 1

Page 7

manufacturing [ˌmænjuˈfæktʃərɪŋ] Herstellung, Produktion
supplier [səˈplaɪə] Anbieter, Lieferant/in
semi-finished [ˌsemiˈfɪnɪʃt] Halbfertig-
plant [plɑːnt] Betrieb, (Industrie-)Anlage, Werk
replace [rɪˈpleɪs] ersetzen, austauschen
reduce [rɪˈdjuːs] reduzieren, verringern
representative [ˌreprɪˈzentətɪv] (Handels-) Vertreter/in
manufacturer [ˌmænjuˈfæktʃərə] Hersteller/in, Produzent/in, Fabrikant
range [reɪndʒ] Sortiment, Auswahl
enclose [ɪnˈkləʊz] beilegen, beifügen

Page 8

management [ˈmænɪdʒmənt] Management, Geschäftsführung
(in)formal [ɪnˈfɔːml] (in)offiziell, (in)formell

Page 9

jumbled-up [ˌdʒʌmbld ˈʌp] durcheinander (gewürfelt)
restore [rɪˈstɔː] restaurieren, wiederherstellen
prompt(ly) [ˈprɒmptli] unverzüglich, sofort, pünktlich
etc [ˌet ˈsetərə] usw.

Page 10

emphasis [ˈemfəsɪs] Betonung, Nachdruck
facility [fəˈsɪləti] Einrichtung
efficiency [ɪˈfɪʃnsi] Effizienz, Leistungsfähigkeit
briefly [ˈbriːfli] kurz

Page 11

profile [ˈprəʊfaɪl] Profil, Porträt, Beschreibung
promotional [prəˈməʊʃənl] Werbe-
equip [ɪˈkwɪp] ausrüsten, ausstatten
electric [ɪˈlektrɪk] elektrisch
throughout [θruːˈaʊt] überall (in)
involved in [ɪnˈvɒlvd ɪn] beteiligt an
expansion [ɪkˈspænʃn] Ausdehnung, Erweiterung, Expansion
research [rɪˈsɜːtʃ] Forschung
reliability [rɪˌlaɪəˈbɪləti] Zuverlässigkeit
expand [ɪkˈspænd] erweitern, expandieren
wherever [weərˈevə] wo (immer)
purchase [ˈpɜːtʃəs] Kauf, Einkaufs-
issue [ˈɪʃuː] Frage, Angelegenheit, Problem, Thema
launch [lɔːntʃ] einführen, starten, auf den Markt bringen
expertise [ˌekspɜːˈtiːz] Sachkenntnis
innovation [ˌɪnəˈveɪʃn] Neuerung, Innovation
investment [ɪnˈvestmənt] Investition, Anlage

Page 12

associate [əˈsəʊʃieɪt] assoziieren, verbinden
indicate [ˈɪndɪkeɪt] anzeigen
location [ləʊˈkeɪʃn] Standort, Lage
found [faʊnd] gründen, begründen
domestic [dəˈmestɪk] für das Inland (gedacht), Inlands-
appliance [əˈplaɪəns] Gerät
workforce [ˈwɜːkfɔːs] Belegschaft, Angestellte
prospect [ˈprɒspekt] Aussicht
objective [əbˈdʒektɪv] Ziel
reduction [rɪˈdʌkʃn] Verringerung
restructure [ˌriːˈstrʌktʃə] umstrukturieren
decision-making [dɪˈsɪʒn meɪkɪŋ] Treffen von Entscheidungen
lean [liːn] schlank

Page 13

term [tɜːm] Ausdruck, Bezeichnung
job-shop [ˈdʒɒbʃɒp] Werkstattfertigung
flow-job [ˈfləʊ dʒɒb] Fließfertigung
cellular [ˈseljʊlə] Zelle(n)-
robotic [rəʊˈbɒtɪk] Roboter-

Page 14

commonly [ˈkɒmənli] häufig, gewöhnlich, üblicherweise
grinding machine [ˈgraɪndɪŋ məʃiːn] Schleifmaschine
heat treat [ˈhiːt triːt] warmbearbeiten, vergüten
assemble [əˈsembl] zusammensetzen, zusammenbauen
irrelevant [ɪˈreləvənt] unwesentlich, nebensächlich
handshake [ˈhændʃeɪk] Händedruck
vary [ˈveəri] unterschiedlich sein, wechseln, sich unterscheiden, variieren
later [ˈleɪtə] später
occasional(ly) [əˈkeɪʒənəli] gelegentlich

Unit 2

Page 15

renovate [ˈrenəveɪt] renovieren
specialize in [ˈspeʃəlaɪz ɪn] sich spezialisieren auf
import [ɪmˈpɔːt] einführen, importieren
export [ɪkˈspɔːt] ausführen, exportieren
accessories [əkˈsesəriz] Zubehör(teile)
renovation [ˌrenəˈveɪʃn] Renovierung
preliminary inquiry [prɪˌlɪmɪnəri ɪnˈkwaɪri] Voruntersuchung
tear out [ˌteər ˈaʊt] herausreißen
reliable [rɪˈlaɪəbl] zuverlässig

Page 16

alteration [ˌɔːltəˈreɪʃn] Änderung
improvement [ɪmˈpruːvmənt] Verbesserung
consumption [kənˈsʌmpʃn] Konsum, Verbrauch
illustrated by [ˈɪləstreɪtɪd baɪ] veranschaulicht durch
plumbing [ˈplʌmɪŋ] sanitär(e Anlagen), Installationen
glazing [ˈgleɪzɪŋ] Verglasung
mahogany [məˈhɒgəni] Mahagoni
conservatory [kənˈsɜːvətri] Gewächshaus
replacement [rɪˈpleɪsmənt] Ersatz, Austausch, Ersatzteil
quotation [kwəʊˈteɪʃn] (Kosten-)Voranschlag, (Preis-)Angebot
(felt) roofing [felt ˈruːfɪŋ] Dachdeckerarbeiten (mit Dachpappe)
Ltd [ˈlɪmɪtɪd] mit beschränkter Haftung, Gesellschaft mit beschränkter Haftung, GmbH
undertake [ˌʌndəˈteɪk] übernehmen
speciality [ˌspeʃiˈæləti] Spezialität, Spezialgebiet
estimate [ˈestɪmət] Schätzung, (Kosten-) Anschlag
outcry [ˈaʊtkraɪ] Aufschrei
dustbin [ˈdʌstbɪn] Mülltonne
throw away [ˌθrəʊ əˈweɪ] wegwerfen
energy source [ˈenədʒi sɔːs] Energiequelle
irreplaceable [ˌɪrɪˈpleɪsəbl] unersetzlich
wasteful [ˈweɪstfl] verschwenderisch
excess [ˈekses] übermäßig, im Übermaß
emission [ɪˈmɪʃn] Emission, Abgabe, Ausstoß
noxious [ˈnɒkʃəs] schädlich
(oil-)fired [ˈɔɪlfaɪəd] Heizöl-
expenditure [ɪkˈspendɪtʃə] Ausgabe
advertising feature [ˌædvətaɪzɪŋ ˈfiːtʃə] Werbeargument
sensibly [ˈsensəbli] vernünftig, sinnvoll
obtain [əbˈteɪn] erhalten, erzielen
benefit [ˈbenɪfɪt] Nutzen, Vorteil

Page 17

ensure [ɪnˈʃʊə] sicherstellen, gewährleisten
proper(ly) [ˈprɒpəli] richtig, ordnungsgemäß
insulate [ˈɪnsjuleɪt] isolieren
gradual(ly) [ˈgrædʒuəli] allmählich, nach und nach
insulation [ˌɪnsjuˈleɪʃn] Isolierung, Isolation
effective(ly) [ɪˈfektɪvli] effektiv, wirksam, wirkungsvoll
retain [rɪˈteɪn] zurückhalten, (be)halten
badly fitting [ˌbædli ˈfɪtɪŋ] schlecht schließend
draught stripping [ˌdrɑːft ˈstrɪpɪŋ] Isolierband (gegen Luftzug)
draught [drɑːft] Zugluft
loft [lɒft] (Dach-)Boden
run up [ˌrʌn ˈʌp] hinauffahren, sich erstrecken
foam [fəʊm] Schaumstoff
semi-detached house [ˌsemi dɪˈtætʃt haʊs] Doppelhaushälfte
cavity [ˈkævəti] Hohlraum
registered with [ˈredʒɪstəd wɪð] registriert bei, eingetragen bei
approve [əˈpruːv] gutheißen, billigen, genehmigen
conserve [kənˈsɜːv] konservieren, erhalten
resource [rɪˈsɔːs] Mittel, Reserve
efficient [ɪˈfɪʃnt] effizient, (leistungs)fähig, tüchtig
fitter [ˈfɪtə] Monteur/in, (Maschinen-) Schlosser/in, Installateur/in
breakthrough [ˈbreɪkθruː] Durchbruch
double glazed [ˌdʌbl ˈgleɪzɪŋ] mit Isolierverglasung
superior [suːˈpɪəriə] hochwertig, hervorragend

thermal ['θɜ:ml] thermisch, Wärme-
key locking ['ki: lɒkɪŋ] abschließbar
espagnolette handle [ɪˌspænjəlet 'hændl] Drehriegel
aluminium [ˌæljə'mɪniəm] Aluminium
reinforcing bar [ˌri:ɪnfɔ:sɪŋ 'bɑ:] Verstärkungsstange
security bead [sɪ'kjʊərəti bi:d] Sicherheitsrandversteifung
vent [vent] Lüftung
available in [ə'veɪləbl ɪn] erhältlich in, verfügbar in
woodgrain ['wʊdgreɪn] Holzmaserung
conform to [kən'fɔ:m tə] entsprechen
workmanship ['wɜ:kmənʃɪp] Arbeit(squalität)
survey ['sɜ:veɪ] Gutachten, Untersuchung
DIY enthusiast [ˌdi: aɪ 'waɪ ɪn'θju:ziæst] Heimwerker
paving ['peɪvɪŋ] Fliesen
cement [sɪ'ment] Zement
brick [brɪk] Ziegelstein, Backstein
hardware ['hɑ:dweə] Haushaltswaren, Eisenwaren
decorate ['dekəreɪt] tapezieren, renovieren
painter & decorator [ˌpeɪntər ənd 'dekəreɪtə] Maler und Dekorateur
re-cover [ˌri:'kʌvə] neu furnieren, neu beziehen
craftsman ['krɑ:ftsmən] (Kunst-) Handwerker/in
pattern ['pætn] Muster
car interior [ˌkɑ:r ɪn'tɪəriə] Wageninneres
trim [trɪm] Innenausstattung
plc [ˌpi: el 'si:] (an der Börse notierte) Aktiengesellschaft, AG
foundation [faʊn'deɪʃn] Stiftung, Anstalt
leak [li:k] undichte Stelle, Leck
burst [bɜ:st] (Wasser-)Rohrbruch
callout service [ˌkɔ:laʊt 'sɜ:vɪs] Notdienst
boiler ['bɔɪlə] Heißwasserspeicher, (Heiz-) Kessel
installer [ɪn'stɔ:lə] Installateur
scheme [ski:m] Plan, Programm

Page 18
incorrect(ly) [ˌɪnkə'rektli] falsch, unrichtig
select [sɪ'lekt] auswählen
blank [blæŋk] Lücke
isolate ['aɪsəleɪt] isolieren
sensible ['sensəbl] vernünftig
sensitive ['sensətɪv] empfindlich, sensibel
surroundings [sə'raʊndɪŋz] Umgebung
contain [kən'teɪn] enthalten
client ['klaɪənt] Kunde, Kundin
bacteria [bæk'tɪəriə] Bakterien
recycling [ˌri:'saɪklɪŋ] Recycling, Wiederverwertung
vessel ['vesl] Gefäß, Behälter
braking fluid [ˌbreɪkɪŋ 'flu:ɪd] Bremsflüssigkeit
carbon monoxide [ˌkɑ:bn mə'nɒksaɪd] Kohlenmonoxid
detect [dɪ'tekt] entdecken

Page 19
equipment [ɪ'kwɪpmənt] Ausrüstung, Ausstattung, Geräte
affect [ə'fekt] beeinflussen, sich auswirken auf

Page 20
extension number [ɪk'stenʃn nʌmbə] Nebennschluss, Direktwahl

Page 22
radiator ['reɪdieɪtə] Heizkörper, Kühler
initial costs [ɪ'nɪʃl] Initialkosten, Gestehungskosten
recommendation [ˌrekəmen'deɪʃn] Empfehlung
annual(ly) ['ænjuəli] jährlich

Page 23
fulfil [fʊl'fɪl] erfüllen
(im)probable [ɪm'prɒbəbl] (un)wahrscheinlich
CFC [ˌsi: ef 'si:] FCKW
ozone layer ['əʊzəʊn leɪə] Ozonschicht
nuclear energy [ˌnju:kliə 'enədʒi] Kernenergie
solar panel [ˌsəʊlə 'pænl] Sonnenkollektor
financially [faɪ'nænʃəli] finanziell, wirtschaftlich
differ ['dɪfə] sich unterscheiden

Unit 3

Page 24
impressed by [ɪm'prest baɪ] beeindruckt durch
assure [ə'ʃʊə] versichern, zusichern
supply [sə'plaɪ] beliefern, liefern, versorgen
Eng. [ɪŋ] Ingenieur
Encl. [ɪn'kləʊʒəz] Anlage(n)

Page 25
commitment [kə'mɪtmənt] Verpflichtung, Engagement

Page 26
optical fibre [ˌɒptɪkl 'faɪbə] Glasfaser
owing to ['əʊɪŋ tə] wegen
rearrange [ˌri:ə'reɪndʒ] umstellen, umändern
schedule ['ʃedju:l] Zeitplan
inconvenience [ˌɪnkən'vi:niəns] Unbequemlichkeit
propose [prə'pəʊz] vorschlagen

Page 27
no-parking zone [ˌnəʊ'pɑ:kɪŋ zəʊn] Parkverbotszone
reservation [ˌrezə'veɪʃn] Reservierung, Buchung

Page 28
bid for ['bɪd fə] ein (An-)Gebot machen
contract ['kɒntrækt] Vertrag, Kontrakt
appropriate [ə'prəʊpriət] geeignet, passend
concrete ['kɒŋkri:t] konkret
durable ['djʊərəbl] dauerhaft, haltbar
unique [ju'ni:k] einzigartig
reputation [ˌrepju'teɪʃn] Ruf
provider [prə'vaɪdə] Lieferant
feature ['fi:tʃə] Merkmal, Kennzeichen
reluctance [rɪ'lʌktəns] magnetischer Widerstand
magnetics [mæg'netɪks] Wissenschaft vom Magnetismus
extensive [ɪk'stensɪv] umfassend
engineering know-how [ˌendʒɪnɪərɪŋ 'nəʊhaʊ] technischer Sachverstand
outstanding [aʊt'stændɪŋ] hervorragend
(high) performance [ˌhaɪ pə'fɔ:məns] Hochleistungs-
load [ləʊd] Ladung, Last, Fracht
steady-state ['stedi steɪt] Ruheposition
stability [stə'bɪləti] Stabilität
repeatability [rɪˌpi:tə'bɪləti] Wiederholbarkeit
maintain [meɪn'teɪn] aufrechterhalten, bewahren
fan [fæn] Gebläse, Ventilator
ventilate ['ventɪleɪt] ventilieren, (be)lüften
hoseproofed ['həʊzpru:ft] schlauchisoliert
maintenance ['meɪntənəns] Wartung, Pflege

Page 29
advanced [əd'vɑ:nst] fortschrittlich
access ['ækses] Zugang, Zugriff
fast-acting [ˌfɑ:st 'æktɪŋ] schnell arbeitend
(forward/reverse) switching [ˌfɔ:wəd rɪˌvɜ:s 'swɪtʃɪŋ] Umschalten (von vorwärts auf rückwärts)
illustrate ['ɪləstreɪt] illustrieren, veranschaulichen
massively ['mæsɪvli] massiv
appoint [ə'pɔɪnt] einstellen, ernennen, berufen
output ['aʊtpʊt] Produktion(smenge), Ausstoß
combustion [kəm'bʌstʃn] Verbrennung
induction motor [ɪn'dʌkʃn məʊtə] Induktionsmotor
trial ['traɪəl] Test, Versuch
comprise [kəm'praɪz] umfassen, bestehen aus
specification enclosure [ˌspesɪfɪ'keɪʃn ɪn'kləʊʒə] beigefügter Bauplan
substantial(ly) [səb'stænʃəli] wesentlich, beträchtlich
convert [kən'vɜ:t] umwandeln, umbauen
extend [ɪk'stend] erweitern, ausdehnen, vergrößern
ability [ə'bɪləti] Fähigkeit
wear and tear [ˌweər ən 'teə] Abnutzung, Verschleiß
drive transmission [ˌdraɪv træns'mɪʃn] Triebwelle
subsidiary [səb'sɪdiəri] Tochtergesellschaft
cooperate [kəʊ'ɒpəreɪt] zusammenarbeiten
mainly ['meɪnli] hauptsächlich
impressive [ɪm'presɪv] beeindruckend
replaced by [rɪ'pleɪst baɪ] ersetzt durch

Page 30
requirement [rɪ'kwaɪəmənt] Bedingung, Bedürfnis
extract ['ekstrækt] Auszug
acceleration [əkˌselə'reɪʃn] Beschleunigung
thus [ðʌs] so(mit)
snag [snæg] Schwierigkeit

Page 31
bolt [bəʊlt] Bolzen, Schraube
fasten ['fɑ:sn] befestigen, anziehen
sleeve [sli:v] Hülse, Büchse, Muffe, Buchse, Manschette
fragile ['frædʒaɪl] zerbrechlich
conventional [kən'venʃənl] herkömmlich
negligible ['neglɪdʒəbl] geringfügig

Page 32
fluctuation [ˌflʌktʃʊ'eɪʃn] Schwankung, Ausschlag
peak [piːk] Gipfel, Spitze(nzeit)
require [rɪ'kwaɪə] benötigen, erfordern
sophisticated [sə'fɪstɪkeɪtɪd] hoch entwickelt, raffiniert
simultaneous(ly) [ˌsɪml'teɪniəsli] gleichzeitig
chart [tʃɑːt] Diagramm, Tabelle
tend to ['tend tə] neigen zu
steadily ['stedɪli] stetig, regelmäßig
shortly before [ˌʃɔːtli bɪ'fɔː] kurz vor
sharply ['ʃɑːpli] rasch, stark
fluctuate ['flʌktʃueɪt] schwanken

Page 33
pattern ['pætn] Muster
anticipate [æn'tɪsɪpeɪt] voraussehen
accordingly [ə'kɔːdɪŋli] (dem)entsprechend
get at ['get ət] hinauswollen auf
instructor [ɪn'strʌktə] Lehrer/in, Ausbilder/in
soften ['sɒfn] abschwächen
cooperative [kəʊ'ɒpərətɪv] kooperativ
flatten out [ˌflætn 'aʊt] (sich) abflachen
recover [rɪ'kʌvə] sich erholen
recovery [rɪ'kʌvəri] Erholung, Aufschwung
plunge [plʌndʒ] abstürzen, Absturz
modifier ['mɒdɪfaɪə] Bestimmungswort
drastically ['dræstɪkli] drastisch
according to [ə'kɔːdɪŋ tə] entsprechend, nach
axis ['æksɪs] Achse
represent [ˌreprɪ'zent] repräsentieren, darstellen
scale [skeɪl] Skala, Maßstab

Page 34
variation [ˌveəri'eɪʃn] Variation, Abweichung

Unit 4

Page 35
bid [bɪd] (An-)Gebot
railway terminal ['reɪlweɪ tɜːmɪnl] Endbahnhof
including (incl.) [ɪn'kluːdɪŋ] einschließlich, inbegriffen
stopover ['stɒpəʊvə] Zwischenlandung, Reiseunterbrechung, kurzer Aufenthalt
celebrate ['selɪbreɪt] feiern
rewrite [ˌriː'raɪt] umschreiben

Page 36
residence ['rezɪdəns] Wohnanlage
gently sloping [ˌdʒentli 'sləʊpɪŋ] sanft abfallend
equal(ly) ['iːkwəli] gleich(ermaßen)
integral part ['ɪntɪgrəl pɑːt] fester Bestandteil
commission [kə'mɪʃn] Kommission, Auftrag
ambitious [æm'bɪʃəs] ehrgeizig
site [saɪt] (Bau-)Stelle, Platz, Gelände
governed by ['gʌvnd baɪ] verwaltet von, geregelt durch
building code ['bɪldɪŋ kəʊd] Bauvorschriften
construction [kən'strʌkʃn] Bau, Konstruktion
mock-up ['mɒkʌp] Modell in natürlicher Größe
(south)-facing ['saʊθ feɪsɪŋ] nach (Süden) liegend
facade [fə'sɑːd] Fassade
echo ['ekəʊ] wieder aufnehmen
wing [wɪŋ] Flügel
raised [reɪzd] erhöht, erhaben
accommodate [ə'kɒmədeɪt] unterbringen, beherbergen
hinge [hɪndʒ] Scharnier
relative(ly) ['relətɪvli] relativ, verhältnismäßig
striking ['straɪkɪŋ] bemerkenswert, eindrucksvoll
dominate ['dɒmɪneɪt] beherrschen

Page 37
association [əˌsəʊsi'eɪʃn] Verband, Gesellschaft
consultant [kən'sʌltənt] Berater/in, Gutachter/in
journal ['dʒɜːnl] Zeitschrift
immediate [ɪ'miːdiət] sofortige(r,s), umgehend
integrate ['ɪntɪgreɪt] eingliedern, integrieren
primarily [praɪ'merəli] vor allem, hauptsächlich
public sector [ˌpʌblɪk 'sektə] öffentliche Hand, öffentlicher Sektor
involvement [ɪn'vɒlvmənt] Beteiligung
constraint [kən'streɪnt] Einschränkung, Beschränkung
look forward to [ˌlʊk 'fɔːwəd tə] sich freuen auf

Page 38
regarding [rɪ'gɑːdɪŋ] betreffs
reference number ['refərəns nʌmbə] Aktenzeichen, Kennziffer
addressee [ˌædre'siː] Empfänger/in, Adressat/in

Page 39
convenient [kən'viːniənt] passend, günstig, bequem

Page 40
on display [ɒn dɪ'spleɪ] ausgestellt
layout ['leɪaʊt] Layout, Grundriss

Page 41
influential [ˌɪnflu'enʃl] einflussreich
distinguish [dɪ'stɪŋgwɪʃ] unterscheiden
distinction [dɪ'stɪŋkʃn] Unterschied, Unterscheidung
precision [prɪ'sɪʒn] Genauigkeit
poetics [pəʊ'etɪks] Kunst
phenomenon [fə'nɒmɪnən] Phänomen
mathematical model [ˌmæθə'mætɪkl mɒdl] mathematisches Modell
interpret [ɪn'tɜːprɪt] interpretieren
purely ['pjʊəli] ausschließlich
visual ['vɪʒuəl] visuell
stress [stres] betonen, hervorheben
contribution [ˌkɒntrɪ'bjuːʃn] Beitrag

Page 42
emancipation [ɪˌmænsɪ'peɪʃn] Emanzipation
attitude ['ætɪtjuːd] Einstellung, Haltung
liberate ['lɪbəreɪt] befreien
time-consuming ['taɪmkənsjuːmɪŋ] zeitraubend
ingenious [ɪn'dʒiːniəs] klug (erdacht), genial
conducted by [kən'dʌktɪd baɪ] durchgeführt von
authority [ɔː'θɒrəti] Verwaltung, Behörde, Amt
responsibility [rɪˌspɒnsə'bɪləti] Verantwortlichkeit, Verantwortung
functional ['fʌŋkʃənl] funktional, zweckmäßig
institutional [ˌɪnstɪ'tjuːʃənl] institutionell
determined by [dɪ'tɜːmɪnd baɪ] bestimmt von
(cold) calculation [kəʊld ˌkælkju'leɪʃn] (kühle) Berechnung
obstacle ['ɒbstəkl] Hindernis
masterpiece ['mɑːstəpiːs] Meisterstück, Meisterwerk

Page 43
(in)definite [ɪn'defɪnət] (un)bestimmt
creativity [ˌkriːeɪ'tɪvəti] Kreativität
qualified by ['kwɒlɪfaɪd baɪ] qualifiziert durch, bestimmt durch
refer to [rɪ'fɜː tə] sich beziehen auf, verweisen auf

Page 44
essential [ɪ'senʃl] (absolut) notwendig, wichtig, wesentlich, (unbedingt) erforderlich
emphasize ['emfəsaɪz] betonen, hervorheben
spokesperson ['spəʊkspɜːsn] Sprecher/in
etiquette ['etɪket] Etikette
candy ['kændi] Bonbon, Praline, Süßigkeiten

Unit 5

Page 45
autotechnology [ˌɔːtətek'nɒlədʒi] Kfz-Technik
automotive [ˌɔːtə'məʊtɪv] Auto-
environmental(ly) [ɪnˌvaɪrən'mentəli] Umwelt-
legislation [ˌledʒɪs'leɪʃn] Gesetzgebung, Gesetze
virtually ['vɜːtʃuəli] praktisch, nahezu
mechanical engineer [mɪˌkænɪkl ˌendʒɪ'nɪə] Maschinen(bau)techniker/in
technological [ˌteknə'lɒdʒɪkl] technologisch
advance [əd'vɑːns] Fortschritt
multi-(national) [ˌmʌlti'næʃnəl] multi(national)
exhibit [ɪg'zɪbɪt] ausstellen
state-of-the-art [ˌsteɪt əv ði 'ɑːt] Stand der Technik
exhibition [ˌeksɪ'bɪʃn] Ausstellung
division [dɪ'vɪʒn] Abteilung

Page 47
prediction [prɪ'dɪkʃn] Vorhersage
arrangement [ə'reɪndʒmənt] Absprache, Vereinbarung
intention [ɪn'tenʃn] Absicht
fall [fɔːl] Herbst
timing ['taɪmɪŋ] Timing, Zeitplanung

Page 48
apparently [ə'pærəntli] offensichtlich, anscheinend
confirm [kən'fɜ:m] bestätigen

Page 49
dismantle [dɪs'mæntl] abwracken, zerlegen, demontieren
achieve [ə'tʃi:v] erreichen, erlangen
dealership ['di:ləʃɪp] Vertretung
drain [dreɪn] absaugen, ablaufen lassen
fluid ['flu:ɪd] Flüssigkeit
contamination [kən,tæmɪ'neɪʃn] Verseuchung, Kontaminierung, Verunreinigung
remove [rɪ'mu:v] entfernen, beseitigen
interior [ɪn'tɪəriə] Inneres, Innenausstattung
segregate ['segrɪgeɪt] trennen
subassembly [,sʌbə'sembli] Untergruppe
bumper ['bʌmpə] Stoßstange

Page 50
component [kəm'pəʊnənt] Bauteil, Bestandteil
gearbox ['gɪəbɒks] Getriebe
axle ['æksl] Achse
starter ['stɑ:tə] Anlasser
wiring ['waɪərɪŋ] Verdrahtung
strip out [,strɪp 'aʊt] demontieren, herausnehmen
bare [beə] nackt, bloß
body shell ['bɒdi ʃel] Karosserie
recondition [,ri:kən'dɪʃn] überholen, erneuern
rebuild [,ri:'bɪld] wiederherstellen
resale ['ri:seɪl] Wiederverkauf, Weiterverkauf
shred [ʃred] shreddern, zerkleinern
(non-)ferrous [,nɒn 'ferəs] (Nicht-)Eisen-
pulp [pʌlp] zerkleinertes Erz
smelt [smelt] (ein)schmelzen, verhütten
reuse [,ri:'ju:s] Wiederverwendung
residue ['rezɪdju:] Rückstand, Rest
landfill ['lændfɪl] geordnete Mülldeponie
calorific value [,kælə,rɪfɪk 'vælju:] Heizwert
initiative [ɪ'nɪʃətɪv] Initiative
potential [pə'tenʃl] möglich, potenziell
considerable (ably) [kən'sɪdərəbli] beträchtlich
dashboard ['dæʃbɔ:d] Armaturenbrett
clarify ['klærəfaɪ] (auf)klären

Page 51
speculate ['spekjuleɪt] spekulieren
suspect [sə'spekt] vermuten
distribute [dɪ'strɪbju:t] verteilen
in-house [,ɪn 'haʊs] innerbetrieblich
in contrast [ɪn 'kɒntrɑ:st] im Gegensatz (dazu)
reconsider [,ri:kən'sɪdə] überdenken
claim [kleɪm] beanspruchen, behaupten
adjust [ə'dʒʌst] anpassen, einstellen
consumer [kən'sju:mə] Verbraucher/in, Konsument/in, Konsum-
scrapyard ['skræpjɑ:d] Schrottplatz
copper ['kɒpə] Kupfer
accidental [,æksɪ'dentl] zufällig
strategy ['strætədʒi] Strategie
refill-container [,ri:fɪl kən'teɪnə] Nachfüllpackung
commercial(ly) [kə'mɜ:ʃəli] kommerziell, wirtschaftlich

Page 52
observe [əb'zɜ:v] beachten
formulate ['fɔ:mjuleɪt] formulieren, ausarbeiten
sequencing ['si:kwənsɪŋ] Reihenfolge, Anordnung
last but not least [,lɑ:st bʌt nɒt 'li:st] nicht zuletzt
sum up [,sʌm 'ʌp] zusammenfassen

Page 53
(un)acceptable [,ʌnək'septəbl] (un)annehmbar, (in)akzeptabel
casual ['kæʒuəl] leger, sportlich
touchy ['tʌtʃi] empfindlich, heikel, kritisch
criticism ['krɪtɪsɪzəm] Kritik
relating to [rɪ'leɪtɪŋ tə] in Bezug auf, in Zusammenhang mit
economy [ɪ'kɒnəmi] Wirtschaft

Unit 6

Page 55
rewards package [rɪ'wɔ:dz pækɪdʒ] Gesamtheit der Vergütungen
leadership ['li:dəʃɪp] Führungsqualität
come out of the blue [,kʌm aʊt əf ðə 'blu:] aus heiterem Himmel kommen
assume [ə'sju:m] annehmen, übernehmen, glauben
prominent ['prɒmɪnənt] prominent, hervorstechend
promotion [prə'məʊʃn] Beförderung

Page 56
land lines ['lænd laɪnz] Festnetz
specifically [spə'sɪfɪkli] speziell, bestimmt
stronger presence [,strɒŋgə 'prezns] stärkere Präsenz
respective [rɪ'spektɪv] jeweilig
equivalent [ɪ'kwɪvələnt] entsprechend
headquarters [,hed'kwɔ:təz] Hauptquartier, Zentrale

Page 57
nappy ['næpi] Windel
frustrated [frʌ'streɪtɪd] frustriert

Page 58
interface ['ɪntəfeɪs] Schnittstelle
participate in [pɑ:'tɪsɪpeɪt ɪn] teilnehmen an
multipurpose [,mʌlti'pɜ:pəs] Vielzweck-
prior to ['praɪə tə] vor
pursue [pə'sju:] verfolgen, nachgehen
laboratory [lə'bɒrətri] Labor(atorium)
former ['fɔ:mə] ehemalig, früher

Page 59
apprenticeship [ə'prentɪʃɪp] Lehre, Ausbildung
employ [ɪm'plɔɪ] anstellen, beschäftigen, einstellen

Page 60
marital status [,mærɪtl 'steɪtəs] Familienstand
preparatory [prɪ'pærətri] vorbereitend
City and Guilds [,sɪti ənd 'gɪldz] Eine Prüfungskammer, die eigene Industrie/Handwerksprüfungen durchführt.
fluent ['flu:ənt] fließend

Page 61
inquire about [ɪn'kwaɪər əbaʊt] sich erkundigen nach, fragen nach

Page 62
fascinating ['fæsɪneɪtɪŋ] faszinierend, hochinteressant
revolutionize [,revə'lu:ʃənaɪz] revolutionieren
rely on [rɪ'laɪ ɒn] sich stützen auf
centrally ['sentrəli] zentral
transmitter [træns'mɪtə] Sender
area of coverage [,eəriə əv 'kʌvərɪdʒ] Reichweite
territorial subdivision [,terə'tɔ:riəl ,sʌbdɪ'vɪʒn] territoriale Unterteilung
frequency ['fri:kwənsi] Frequenz
allocate ['æləkeɪt] zuweisen, zuteilen
subscriber [səb'skraɪbə] Abonnent/in
adjoining [ə'dʒɔɪnɪŋ] anliegend, benachbart
fraction ['frækʃn] Bruchteil
traverse [trə'vɜ:s] durchqueren
eventually [ɪ'ventʃuəli] schließlich, zum Schluss
adjacent [ə'dʒeɪsnt] angrenzend, benachbart
handing-off [,hændɪŋ 'ɒf] Übergabe-
transfer [træns'fɜ:] übermitteln, übertragen
detector [dɪ'tektə] Detektor

Page 63
determine [dɪ'tɜ:mɪn] bestimmen, feststellen
cell site ['sel saɪt] Zelle/Zellenstandort
confer [kən'fɜ:] sich abstimmen
frequency pair ['fri:kwənsi peə] Frequenzpaar
transmit [træns'mɪt] übertragen
instruction [ɪn'strʌkʃn] Anweisung
voltage ['vəʊltɪdʒ] Spannung
frequency synthesizer [,fri:kwənsi 'sɪnθəsaɪzə] Frequenzsynthesizer
unnoticeable [,ʌn'nəʊtɪsəbl] unmerklich
millisecond ['mɪlisekənd] Millisekunde
monitor ['mɒnɪtə] überwachen
adjustment [ə'dʒʌstmənt] Einstellung, Anpassung
cellular telephony [,seljələ tə'lefəni] Mobilfunk
consequently ['kɒnsɪkwəntli] dementsprechend, infolgedessen
predetermined [,pri:dɪ'tɜ:mɪnd] vorherbestimmt
conversely [kən'vɜ:sli] umgekehrt
instruct [ɪn'strʌkt] anweisen
discrete [dɪ'skri:t] getrennt, einzeln
interfere [,ɪntə'fɪə] stören
interference [,ɪntə'fɪərəns] Störung

Page 64
column ['kɒləm] Spalte, Kolumne
precisely [prɪ'saɪsli] genau
disregard [,dɪsrɪ'gɑ:d] ignorieren, nicht beachten
immobile [ɪ'məʊbaɪl] unbeweglich
character trait ['kærəktə treɪ] Charakterzug
communicator [kə'mju:nɪkeɪtə] Unterhalter/in
harmonious [hɑ:'məʊniəs] harmonisch
group mentality [,gru:p men'tæləti] Gruppenmentalität

Module 1

Page 66

drawing ['drɔ:ɪŋ] Ziehen, Walzen
turning ['tɜ:nɪŋ] Drehen, Drechseln
crankshaft ['kræŋkʃɑ:ft] Kurbelwelle
camshaft ['kæmʃɑ:ft] Nockenwelle
consult [kən'sʌlt] konsultieren, zu Rate ziehen, nachschlagen in
mechanism ['mekənɪzəm] Mechanismus
wire drawing ['waɪə drɔ:ɪŋ] Drahtziehen
die [daɪ] (Draht-)Ziehstahl
diameter [daɪ'æmɪtə] Durchmesser
workpiece ['wɜ:kpi:s] Werkstück
operator ['ɒpəreɪtə] Maschinist/in
(pair of) pliers [ˌpeər əv 'plaɪəz] (eine) Zange
flywheel ['flaɪwi:l] Schwungrad
foot treadle ['fʊt tredl] Fußantrieb
running centre ['rʌnɪŋ sentə] Laufzentrum

Page 67

mechanization [ˌmekənaɪ'zeɪʃn] Mechanisierung
medieval [ˌmedi'i:vl] mittelalterlich
attach to [ə'tætʃ tə] befestigen an
mount [maʊnt] montieren
bearings ['beərɪŋz] Lager
device [dɪ'vaɪs] Vorrichtung, Gerät
fastened to ['fɑ:snd tə] befestigt an
in such a manner [ɪn ˌsʌtʃ ə 'mænə] so
jaws [dʒɔ:z] Backen, Maul
grip [grɪp] (er)greifen
crank [kræŋk] Kurbel
lengthen ['leŋθən] verlängern
release [rɪ'li:s] freilassen, freigeben
slide back [ˌslaɪd 'bæk] zurückgleiten
thread [θred] einfädeln, (hindurch)ziehen
sharpen ['ʃɑ:pən] schärfen
friction ['frɪkʃn] Reibung
lubrication [ˌlu:brɪ'keɪʃn] Schmierung

Page 68

oscillating [ˌɒsɪ'leɪtɪŋ] oszillierend
permit [pə'mɪt] erlauben
enable [ɪ'neɪbl] ermöglichen
reformulate [ˌri:'fɔ:mjuleɪt] umformulieren
transform [træns'fɔ:m] verwandeln
reciprocating [rɪ'sɪprəkeɪtɪŋ] sich hin und her bewegend
transformation [ˌtrænzfə'meɪʃn] Umwandlung
position [pə'zɪʃn] lokalisieren, anordnen
flexibly ['fleksəbli] flexibel
lateral ['lætərəl] seitlich
self-enforcing [ˌself ɪn'fɔ:sɪŋ] selbsttätig
suspend [sə'spend] aufhängen
(anti)clockwise [ˌænti'klɒkwaɪz] im/gegen den Uhrzeigersinn
intermittent [ˌɪntə'mɪtənt] mit Unterbrechungen

Page 69

classify ['klæsɪfaɪ] klassifizieren, einteilen
category ['kætəgəri] Kategorie, Klasse
restrict [rɪ'strɪkt] einschränken
worktable ['wɜ:kteɪbl] Werktisch
be proportional to [bi prə'pɔ:ʃənl tə] proportional sein zu, im Verhältnis stehen zu
numeric [nju:'merɪk] numerisch
directly [də'rektli] direkt
tactile ['tæktaɪl] tastbar, Tast-
visual sensing [ˌvɪʒuəl 'sensɪŋ] optische Erfassung
optically ['ɒptɪkli] optisch

Page 70

gripper ['grɪpə] Greifer
piezo electric device [ˌpi:zəʊ ɪˌlektrɪk dɪ'vaɪs] piezoelektrische Vorrichtung
strain gauge ['streɪn geɪdʒ] Dehnungsmesser
to such an extent [tə ˌsʌtʃ ən ɪk'stent] in solchem Ausmaß
slipping ['slɪpɪŋ] Rutschen
take for granted [ˌteɪk fə 'grɑ:ntɪd] als selbstverständlich ansehen
insert [ɪn'sɜ:t] einfügen
circuit ['sɜ:kɪt] Schaltung
solder joint ['səʊldə dʒɔɪnt] Lötstelle
crack [kræk] Riss, Spalt(e)
imperfection [ˌɪmpə'fekʃn] Unvollkommenheit, Mangel
microprocessor ['maɪkrəʊprəʊsesə] Mikroprozessor
digitize ['dɪdʒɪtaɪz] digitalisieren
inspect [ɪn'spekt] inspizieren
defective [dɪ'fektɪv] fehlerhaft
random(ly) ['rændəmli] zufällig

Page 71

productivity [ˌprɒdʌk'tɪvəti] Produktivität
lubricate ['lu:brɪkeɪt] schmieren

Page 72

servo-mechanism [ˌsɜ:vəʊ 'mekənɪzəm] Servomechanismus
amplify ['æmplɪfaɪ] verstärken
amplification [ˌæmplɪfɪ'keɪʃn] Verstärkung
plastic extruding [ˌplæstɪk ɪk'stru:dɪŋ] Plastikpressen
granule ['grænju:l] Granulat
rod [rɒd] Stab, Stange
hopper ['hɒpə] Trichter
melting ['meltɪŋ] Schmelzen, Schmelz-
melt-pumping zone ['meltpʌmpɪŋ zəʊn] Schmelzdruckbereich

Page 73

sack [sæk] feuern, entlassen, rauswerfen
redesign [ˌri:dɪ'zaɪn] umstrukturieren
(sub)assembly line [ˌsʌbə'sembli laɪn] (Neben-)Fließband
spark [spɑ:k] Funke
weld [weld] schweißen
brake lining [ˌbreɪk 'laɪnɪŋ] Bremsbelag
overhead ['əʊvəhed] oben, darüber
track [træk] verfolgen
arguably ['ɑ:gjuəbli] darüber lässt sich streiten
forefront ['fɔ:frʌnt] Spitze
perform [pə'fɔ:m] ausführen, erledigen, durchführen
adviser [əd'vaɪzə] Berater/in
wisdom ['wɪzdəm] Weisheit
march [mɑ:tʃ] Marsch
significant(ly) [sɪg'nɪfɪkəntli] bedeutend, merklich, erheblich
subtract [səb'trækt] abziehen, verringern

Page 74

labor ['leɪbə] Arbeit, Arbeitskräfte, Arbeitskraft
demoralize [dɪ'mɒrəlaɪz] demoralisieren, entmutigen
maintain [meɪn'teɪn] warten
retooling [ˌri:'tu:lɪŋ] Neuausrüstung (mit Werkzeugen oder Maschinen)
autoworker ['ɔ:təwɜ:kə] Kfz-Arbeiter/in
rate [reɪt] Rate, Prozentsatz
roughly ['rʌfli] ungefähr
entire [ɪn'taɪə] ganz
tend [tend] bedienen
segment [seg'ment] unterteilen
steering wheel ['stɪərɪŋ wi:l] Lenkrad, Steuer
micron ['maɪkrɒn] Mikron, Mikrometer
conduct [kən'dʌkt] durchführen
diagnostic [ˌdaɪəg'nɒstɪk] diagnostisch
sour on ['saʊər ɒn] sauer werden in Bezug auf
selective [sɪ'lektɪv] wählerisch
expectation [ˌekspek'teɪʃn] Erwartung

Page 75

concentrate on ['kɒnsntreɪt ɒn] sich konzentrieren auf
make redundant [ˌmeɪk rɪ'dʌndənt] entlassen, freisetzen
partial(ly) ['pɑ:ʃəli] teilweise

Page 76

manual(ly) ['mænjuəli] mit der Hand
evidence ['evɪdəns] Beweismaterial, Beweis(e)
phonetic transcription [fəˌnetɪk træn'skrɪpʃn] Lautschrift
syllable ['sɪləbl] Silbe
derogatory [dɪ'rɒgətəri] abfällig, abträglich
critical ['krɪtɪkl] kritisch

Module 2

Page 77

potable ['pəʊtəbl] trinkbar
ecological [ˌi:kə'lɒdʒɪkl] ökologisch
cubic-(foot) ['kju:bɪk fʊt] Kubik(fuß) (1 cu foot = 0.028 m3)
correspond to [ˌkɒrɪ'spɒnd tə] entsprechen
carbon dioxide [ˌkɑ:bən daɪ'ɒksaɪd] Kohlendioxid
release [rɪ'li:s] freisetzen, entlassen
collector [kə'lektə] Sammler, Kollektor
displace [dɪs'pleɪs] ersetzen
loop [lu:p] Schleife

Page 78

pass through ['pɑ:s θru:] durchlaufen, passieren
coil [kɔɪl] Windung, Spule
transfer [træns'fɜ:] verwandeln
(heat) exchanger [hi:t ɪks'tʃeɪndʒə] Wärmetauscher
device [dɪ'vaɪs] Vorrichtung, Gerät, Regler
exceed [ɪk'si:d] übersteigen
valve [vælv] Ventil
drain back ['dreɪn bæk] zurücksaugen, zurückfließen
freeze-up ['fri:z ʌp] Einfrieren
open/closed-loop [ˌəʊpən, ˌkləʊzd 'lu:p] mit offener/geschlossener Schleife
frost-protecting ['frɒst prətektɪŋ] gegen Frost schützen

Page 79

pumped draindown system [ˌpʌmpt ˌdreɪndaʊn ˈsɪstəm] Ein System, in dem das Wasser bei Frostgefahr abgelassen wird und später wieder mit einer Pumpe aufgefüllt wird.
sub-zero [ˌsʌbˈzɪərəʊ] unter Null
piping arrangement [ˌpaɪpɪŋ əˈreɪndʒmənt] Rohrverbindung
overhead transparency [ˌəʊvəhed træns'pærənsi] Tageslichtfolie
despite [dɪˈspaɪt] trotz
mount on [ˈmaʊnt ɒn] montieren auf
slanted [ˈslɑːntɪd] geneigt, schräg
header [ˈhedə] Rohrverbinder
link [lɪŋk] verbinden
riser [ˈraɪzə] Steigleitung, Steigrohr
in such a manner [ɪn ˌsʌtʃ ə ˈmænə] so
joint [dʒɔɪnt] Verbindungsrohr
incorporate [ɪnˈkɔːpəreɪt] enthalten
attached to [əˈtætʃt tə] befestigt an
drain pipe [ˈdreɪnpaɪp] Abflussrohr, Abzugsröhre
auxiliary [ɔːgˈzɪliəri] zusätzlich
gain [geɪn] gewinnen

Page 80

intensity [ɪnˈtensəti] Intensität
tracking [ˈtrækɪŋ] verfolgend
transcript [ˈtrænskrɪpt] Abschrift, Niederschrift
rate [reɪt] Prozentsatz, Rate
ambient [ˈæmbiənt] umgebend
superimpose [ˌsuːpərɪmˈpəʊz] hinzufügen, zur Deckung bringen

Page 81

proportion [prəˈpɔːʃn] Teil, Anteil
input [ˈɪnpʊt] Eingabe
square meter [ˌskweə ˈmiːtə] Quadratmeter
reveal [rɪˈviːl] enthüllen, zeigen
initially [ɪˈnɪʃəli] anfänglich, zuerst
optimum [ˈɒptɪməm] Optimum, optimal
additional [əˈdɪʃənl] zusätzlich
minimal [ˈmɪnɪməl] minimal

Page 82

solely [ˈsəʊlli] einzig und allein
restricted to [rɪˈstrɪktɪd tə] beschränkt auf
anti-freeze [ˈæntifriːz] Gefrier(mittel), Frostschutz(mittel)
hazard [ˈhæzəd] Risiko, Gefahr
toxic [ˈtɒksɪk] giftig, toxisch
tidal [ˈtaɪdl] Gezeiten-
inventor [ɪnˈventə] Erfinder/in
linking word [ˌlɪŋkɪŋ ˈwɜːd] Bindewort
tide [taɪd] Gezeiten
float [fləʊt] Schwimmer
piston [ˈpɪstən] Kolben
inlet [ˈɪnlet] Einlass, Öffnung
snorkel [ˈsnɔːkl] Schnorchel
outlet [ˈaʊtlet] Abfluss, Ausgang
(sub)atmospheric pressure [ˌsʌbˌætməsˈferɪk ˈpreʃə] (Unter-) Luftdruck
chamber [ˈtʃeɪmbə] Kammer

Page 83

differential thermostat [ˌdɪfəˌrenʃl ˈθɜːməstæt] Differentialthermostat
impart [ɪmˈpɑːt] mitteilen, übertragen
blade [bleɪd] Blatt, Schaufel

Page 84

pressurized-water reactor [ˌprəʃəraɪzd ˌwɔːtə riˈæktə] Druckwasserreaktor
uranium dioxide [juˌreɪniəm daɪˈɒksaɪd] Urandioxid
enrich [ɪnˈrɪtʃ] anreichern
hermetically sealed [hɜːˈmetɪkli siːld] hermetisch verschlossen
zirconium [zɜːˈkəʊniəm] Zirkonium
alloy [ˈælɔɪ] Legierung
tubing [ˈtjuːbɪŋ] Röhren
constitute [ˈkɒnstɪtjuːt] darstellen
barrier [ˈbæriə] Schranke
fission [ˈfɪʃn] Spaltung
assembly [əˈsembli] Gruppe
(zirconium-)clad [zɜːˈkəʊniəm klæd] (mit Zirkonium) verkleidet
heavy-walled [ˌhevi ˈwɔːld] dickwandig
neutron [ˈnjuːtrɒn] Neutron
probability [ˌprɒbəˈbɪləti] Wahrscheinlichkeit
to such an extent [tə ˌsʌtʃ ən ɪkˈstent] in solchem Ausmaß
critical [ˈkrɪtɪkl] kritisch
sustain [səˈsteɪn] aufrechterhalten
variable [ˈveəriəbl] veränderlich, regelbar
neutron absorbing boron [ˌnjuːtrɒn əbˌsɔːbɪŋ ˈbɔːrɒn] das Neutronen absorbierende Bor
boric acid [ˌbɔːrɪk ˈæsɪd] Borsäure
dissolve [dɪˈzɒlv] (sich) auflösen
circulating pump [ˌsɜːkjuleɪtɪŋ ˈpʌmp] Umlaufspumpe
gas-cushioned pressurizer [ˌgæskʊʃnd ˈpreʃəraɪzə] Druckausgleichsbehälter

Page 85

drive [draɪv] antreiben
condenser [kənˈdensə] Kondensator
condensate [ˈkɒndenseɪt] Kondensat
recirculate [ˌriːˈsɜːkjəleɪt] umleiten
disposal [dɪˈspəʊzl] Beseitigung, Entsorgung
adverse [ˈædvɜːs] ungünstig, nachteilig, negativ
fossil [ˈfɒsl] fossil
obtainable [əbˈteɪnəbl] erhältlich, zu beziehen
impact [ˈɪmpækt] Wirkung, Einfluss
satisfactorily [ˌsætɪsˈfæktərəli] zufrieden stellend
site on [ˈsaɪt ɒn] platzieren auf, stationieren auf
adequate [ˈædɪkwət] ausreichend, adäquat
absorption [əbˈsɔːpʃn] Aufnahme
capacity [kəˈpæsəti] Kapazität, Leistung, Fähigkeit
omit [əˈmɪt] auslassen, weglassen
sentence switchboard [ˌsentəns ˈswɪtʃbɔːd] Satzbautafel
dissipated [ˈdɪsɪpeɪtɪd] zerstreut, aufgelöst

Page 86

pros and cons [ˌprəʊz ənd ˈkɒnz] Für und Wider
global warming [ˌgləʊbl ˈwɔːmɪŋ] Erderwärmung, Erwärmung der Erdatmosphäre
chairperson [ˈtʃeəpɜːsn] Vorsitzende/r, Moderator/in
raise [reɪz] erhöhen

Page 87

vapour [ˈveɪpə] Dampf, Dunst
agent [ˈeɪdʒənt] Handelnde/r
assembler [əˈsemblə] Monteur
consideration [kənˌsɪdəˈreɪʃn] Erwägung, Überlegung
index card [ˈɪndeks kɑːd] Karteikarte
partition [pɑːˈtɪʃn] Abteilung
interval [ˈɪntəvl] Abstand, Pause

Module 3

Page 88

rectangular frame [rekˌtæŋgjələ ˈfreɪm] rechteckiger Rahmen, Rechteck
magnetic [mægˈnetɪk] magnetisch
distinct [dɪˈstɪŋkt] deutlich, klar erkennbar
transparent [trænsˈpærənt] durchsichtig, durchlässig
damping element [ˌdæmpɪŋ ˈelɪmənt] Dämmelement
polarity [pəˈlærəti] Polarität
reverse [rɪˈvɜːs] umkehren
device [dɪˈvaɪs] Gerät
coil meter [ˈkɔɪl miːtə] Spannungsmesser

Page 90

alternator [ˈɔːltəneɪtə] Wechselstromgenerator
conductor [kənˈdʌktə] Konduktor
induce [ɪnˈdjuːs] auslösen, induzieren
electromotive force (e.m.f) [ɪlektrəʊˌməʊtɪv ˈfɔːs] elektromotorische Kraft
magnitude [ˈmægnɪtjuːd] Größe
pole [pəʊl] Pol
furthermore [ˌfɜːðəˈmɔː] ferner, überdies
clockwise [ˈklɒkwaɪz] im Uhrzeigersinn
permanent [ˈpɜːmənənt] permanent, Dauer-
light bulb [ˈlaɪt bʌlb] Glühbirne
brush [brʌʃ] (Kontakt-)Bürste, Stromabnehmer
rub [rʌb] reiben, schleifen
slip ring [ˈslɪp rɪŋ] Schleifrad
stable [ˈsteɪbl] stabil
right angle [ˈraɪt æŋgl] rechter Winkel
brief [briːf] kurz

Page 91

solder [ˈsəʊldə] löten
winding [ˈwaɪndɪŋ] Windung
counter-clockwise [ˌkaʊntəˈklɒkwaɪz] gegen den Uhrzeigersinn, links herum
intermittent(ly) [ˌɪntəˈmɪtəntli] mit Unterbrechungen
spin [spɪn] (schnell) drehen
mount [maʊnt] montieren
exert [ɪgˈzɜːt] ausüben, anwenden

Page 92

unscrew [ˌʌnˈskruː] abschrauben
plug [plʌg] Stöpsel, Stecker
flex clamp [ˈfleks klæmp] Litzenschelle
cord grip [ˈkɔːd grɪp] Kabelhalter
fuse [fjuːz] Sicherung
lever out [ˈliːvər aʊt] (her)aushebeln
screwdriver [ˈskruːdraɪvə] Schraubenzieher
outer sheath [ˌaʊtə ˈʃiːθ] Ummantelung
tighten [ˈtaɪtn] anziehen
stray whiskers [streɪ ˈwɪskəz] lose Drähtchen

refit [ˌriːˈfɪt] wieder einsetzen
strip [strɪp] entfernen, abisolieren
expose [ɪkˈspəʊz] freilegen
screwhole terminal [ˌskruːhəʊl ˈtɜːmɪnl] Schraubanschluss
strand [strænd] Strang, Strähne
twist together [twɪst təˈgeðə] zusammendrehen
derive from [dɪˈraɪv frəm] ableiten von
grease [griːs] Fett, (ein)fetten
seal [siːl] Siegel, versiegeln

Page 93
socket outlet [ˌsɒkɪt ˈaʊtlet] Steckdose
substitute [ˈsʌbstɪtjuːt] Ersatz-
intact [ɪnˈtækt] intakt
moulded plug [ˌməʊldɪd ˈplʌg] eingeschweisster Stecker
(non) rewirable [ˌnɒn rɪˈwaɪərəbl] (nicht) neu anschließbar
unplug [ˌʌnˈplʌg] den Stecker herausziehen
prise out [ˈpraɪz aʊt] heraushebeln
fuseholder [ˈfjuːzhəʊldə] Sicherungsfassung
dispose of [dɪˈspəʊz əv] beseitigen, entsorgen
consequent [ˈkɒnsɪkwənt] (dem)entsprechend
torch [tɔːtʃ] Taschenlampe
(torch) casing [tɔːtʃ ˈkeɪsɪŋ] (Lampen-) Gehäuse

Page 94
omission [əˈmɪʃn] Auslassen
disassembly [ˌdɪsəˈsembli] Demontage, Zerlegung
manual [ˈmænjuəl] Handbuch
category [ˈkætəgəri] Kategorie, Klasse
align [əˈlaɪn] ausrichten
evaluate [ɪˈvæljueɪt] (ein)schätzen, bewerten
inspect [ɪnˈspekt] inspizieren
lubricate [ˈluːbrɪkeɪt] schmieren
release [rɪˈliːs] freilassen, freigeben
slide down [ˌslaɪd ˈdaʊn] hinabgleiten

Page 95
armature coil [ˈɑːmətʃə kɔɪl] Ankerwicklung
slot [slɒt] Nut(e), Kerbe
cylindrical [səˈlɪndrɪkl] zylindrisch
loop [luːp] Schleife
semicircular [ˌsemiˈsɜːkjələ] halbkreisförmig
segment [ˈsegmənt] Teilstück, Segment
commutator [ˈkɒmjuteɪtə] Kollektor, Stromwender
accomplish [əˈkʌmplɪʃ] schaffen, erreichen
stationary [ˈsteɪʃənri] fest(stehend)
for clarity [fə ˈklærəti] aus Gründen der Klarheit
electrical(ly) [ɪˈlektrɪkli] elektrisch
slight [slaɪt] gering, leicht
revolution [ˌrevəˈluːʃn] Drehung

Page 96
wind around [waɪnd] wickeln um
lateral [ˈlætərəl] seitlich
outward [ˈaʊtwəd] (nach) außen

Module 4

Page 97
feather [ˈfeðə] Feder
appearance [əˈpɪərəns] Äußeres, Aussehen
rib [rɪb] Rippe
quill [kwɪl] Spule, Federkiel
evolve [ɪˈvɒlv] entwickeln, erzeugen
resist [rɪˈzɪst] widerstehen, sich wehren (gegen)
withstand [wɪðˈstænd] widerstehen, Widerstand leisten
span [spæn] umspannen
load [ləʊd] Last, Belastung
static [ˈstætɪk] statisch
extreme [ɪkˈstriːm] äußerst, extrem, außerordentlich
despite [dɪˈspaɪt] trotz
proverbial [prəˈvɜːbiəl] sprichwörtlich
lightness [ˈlaɪtnəs] Leichtigkeit
rectangular [rekˈtæŋgjələ] rechteckig
core [kɔː] Kern
structural [ˈstrʌktʃərəl] strukturell, Bau-, Konstruktions-
simplify [ˈsɪmplɪfaɪ] vereinfachen

Page 98
beam [biːm] (Licht-)Strahl
tension [ˈtenʃn] (An-)Spannung
compression [kəmˈpreʃn] Kompression, Druck
adhesive tape [ədˈhiːsɪv teɪp] Klebeband, Heftpflaster
rod [rɒd] Stab, Stange
grid [grɪd] Gitter
deflect [dɪˈflekt] ablenken, umbiegen, abweichen
occur [əˈkɜː] geschehen, vorkommen
spacing [ˈspeɪsɪŋ] Abstand
roughly [ˈrʌfli] ungefähr
imaginary [ɪˈmædʒɪnəri] imaginär, gedacht
compress [kəmˈpres] komprimieren, zusammenpressen
unstable [ʌnˈsteɪbl] nicht stabil
reinforce [ˌriːɪnˈfɔːs] verstärken
cast [kɑːst] gießen
tensile strength [ˌtensaɪl ˈstreŋθ] Dehnfestigkeit, Zugfestigkeit
compressive [kəmˈpresɪv] Press-, Druck-
solid [ˈsɒlɪd] fest(er Körper)
hollow [ˈhɒləʊ] hohl
cross-section [ˈkrɒs sekʃn] Querschnitt
rigidity [rɪˈdʒɪdəti] Stabilität
box-section [ˈbɒks sekʃn] mit quadratischem Querschnitt
crumble [ˈkrʌmbl] zerfallen, zerbröseln

Page 100
for clarity [fə ˈklærəti] aus Gründen der Klarheit
additional [əˈdɪʃənl] zusätzlich
contractor [kənˈtræktə] Unternehmer (mit Werkvertrag)
precast [ˌpriːˈkɑːst] vorfabrizieren
deck segment [dek ˈsegmənt] Plattformsegment
shore [ʃɔː] Ufer
fleet [fliːt] Flotte
amphibious [æmˈfɪbiəs] amphibisch, Amphibien-
craft [krɑːft] Fahrzeug(e), Schiff(e)
tide [taɪd] Gezeiten, Flut
alternate [ɔːlˈtɜːnət] abwechselnd

pylon [ˈpaɪlən] Mast
casting [ˈkɑːstɪŋ] Guss, Gießen

Page 101
suspension bridge [səˈspenʃn brɪdʒ] Hängebrücke
overview [ˈəʊvəvjuː] Überblick
pick up the thread [pɪk ˌʌp ðə ˈθred] den Faden wieder aufnehmen
arch bridge [ˈɑːtʃ brɪdʒ] Bogenbrücke
subject to [səbˈdʒekt tə] unterwerfen
rivet [ˈrɪvɪt] Niete
clamp together [klæmp təˈgeðə] festklemmen
wedge [wedʒ] Keil
metallurgy [meˈtælədʒi] Metallurgie, Hüttenkunde
refine [rɪˈfaɪn] raffinieren, verfeinern
wrought iron [ˌrɔːt ˈaɪən] Schmiedeeisen, Schweißeisen, Stabeisen
supersede [ˌsuːpəˈsiːd] ersetzen, ablösen
advent [ˈædvent] Ankunft, Aufkommen
truss bridge [ˈtrʌs brɪdʒ] Fachwerkbrücke
cantilever bridge [ˈkæntɪliːvə brɪdʒ] Auslegerbrücke
minimize [ˈmɪnɪmaɪz] verkleinern, verringern
achievement [əˈtʃiːvmənt] Leistung, Errungenschaft
arch [ɑːtʃ] Bogen
cantilever [ˈkæntɪliːvə] aufhängen
pier [pɪə] Pier, Pfeiler
midspan [ˈmɪdspæn] Mittelteil (einer Auslegerbrücke)
counter-balanced [ˌkaʊntəˈbælənst] (durch Gegengewicht) ausgeglichen
temporary [ˈtemprəri] provisorisch, Hilfs-
erect [ɪˈrekt] (er)bauen, errichten
suspended [səˈspendɪd] hängend, schwebend
reveal [rɪˈviːl] enthüllen, zeigen
superstructure [ˈsuːpəstrʌktʃə] Aufbau
Fahrenheit [ˈfærənhaɪt] Fahrenheit
ambient [ˈæmbiənt] umgebend
adjustable [əˈdʒʌstəbl] verstellbar
anticipation [ænˌtɪsɪˈpeɪʃn] Erwartung
restricted to [rɪˈstrɪktɪd tə] beschränkt auf
borne by [ˈbɔːn baɪ] getragen durch

Page 102
content [ˈkɒntent] Gehalt, Inhalt
alloy [ˈælɔɪ] Legierung
galvanized steel [ˌgælvənaɪzd ˈstiːl] verzinkter Stahl
possess [pəˈzes] besitzen
spin together [spɪn təˈgeðə] zusammendrehen
undoubtedly [ʌnˈdaʊtɪdli] zweifellos, ohne (jeden) Zweifel
remarkable [rɪˈmɑːkəbl] bemerkenswert, beachtlich
foundation [faʊnˈdeɪʃn] Grundlage, Fundament
collapse [kəˈlæps] Zusammenbruch, Einsturz
wave-like [ˈweɪvlaɪk] wellenförmig
galloping [ˈgæləpɪŋ] galoppierend
destruction [dɪˈstrʌkʃn] Zerstörung, Vernichtung
inadequate [ɪnˈædɪkwət] nicht ausreichend, ungenügend
aerodynamic [ˌeərəʊdaɪˈnæmɪk] aerodynamisch

airplane ['eəpleɪn] Flugzeug

Page 103

accompanied by [ə'kʌmpənid baɪ] begleitet von
stabilize ['steɪbəlaɪʒ] stabilisieren
telescopic mast [ˌtelɪ'skɒpɪk mɑːst] Teleskopmast
permanent ['pɜːmənənt] ständig, dauerhaft

Page 104

raise [reɪz] aufrichten
vital ['vaɪtl] (lebens)wichtig, wesentlich
joint [dʒɔɪnt] gemeinsam
capacity [kə'pæsəti] Kapazität, Fassungsvermögen, Platz
spectacular [spek'tækjələ] sensationell, spektakulär
domed [dəʊmd] mit einer Kuppel versehen, kuppelförmig
slipformed ['slɪpfɔːmd] Gießverfahren für Beton
slung from [slʌŋ frəm] herabhängen von, aufgehängt an
(steel-)framed [ˌstiːl'freɪmd] Stahlbau-
sequence ['siːkwəns] Reihenfolge
cranage ['kreɪnɪdʒ] Krangebühr
hoist up ['hɔɪst ʌp] hochziehen
jack [dʒæk] Hebevorrichtung
attached to [ə'tætʃt tə] verbunden mit, befestigt an
load-bearing pin ['ləʊdbeərɪŋ pɪn] die Last tragender Bolzen
guy [gaɪ] Halteseil
rigid ['rɪdʒɪd] starr
tensioned ['tenʃnd] Spann-
contract [kən'trækt] zusammenziehen
subcontractor [ˌsʌbkən'træktə] Subunternehmer
slenderness ['slendənəs] Schlankheit
high-rise ['haɪraɪz] Hochhaus-
swaying ['sweɪɪŋ] Schwingen, Schwanken
excessive [ɪk'sesɪv] übermäßig, übertrieben
disrupt [dɪs'rʌpt] unterbrechen
bounce [baʊns] (ab)prallen
deflection movement [dɪ'flekʃn muːvmənt] Ablenkungsbewegung
stiffen ['stɪfn] (ver)steifen, (ver)stärken
fan out [ˌfæn 'aʊt] sich (fächerförmig) ausbreiten, sich erstrecken
tensioning ['tenʃnɪŋ] Spannung
downwind [ˌdaʊn'wɪnd] in Windrichtung
go slack [gəʊ 'slæk] schlaff werden, erschlaffen
tilting pressure [ˌtɪltɪŋ 'preʃə] Kippdruck
exert [ɪg'zɜːt] ausüben, anwenden

Page 105

stable ['steɪbl] stabil
disorder [dɪs'ɔːdə] Störung, Unordnung
sideways ['saɪdweɪz] seitwärts, zur Seite
draft [drɑːft] Entwurf
finalise ['faɪnəlaɪz] endgültig festlegen, abschließen

Module 5

Page 106

water treatment [ˌwɔːtə 'triːtmənt] Wasseraufbereitung
depict [dɪ'pɪkt] darstellen

Page 107

drain down ['dreɪn daʊn] absaugen, abfließen
take for granted [ˌteɪk fə 'grɑːntɪd] als selbstverständlich betrachten
borehole ['bɔːhəʊl] Bohrloch
screen [skriːn] (Gitter-)Sieb, (Draht-)Gitter
floating debris [ˌfləʊtɪŋ 'debriː] schwimmende Teile
storage reservoir [ˌstɔːrɪdʒ 'rezəvwɑː] Speicherbecken
capacity [kə'pæsəti] Kapazität, Fassungsvermögen
plentiful ['plentɪfl] reichlich (vorhanden)
drought [draʊt] Trockenzeit, Dürre(periode)
purification [ˌpjʊərɪfɪ'keɪʃn] Reinigung, Säuberung
tap [tæp] Wasserhahn
stillness ['stɪlnəs] Stille
impurity [ɪm'pjʊərəti] Verunreinigung, Unreinheit
oxygen ['ɒksɪdʒən] Sauerstoff
filtration [fɪl'treɪʃn] Filtern, Filtrierung
coarse [kɔːs] grob
at a rate of [ət ə 'reɪt əv] mit einer Geschwindigkeit von
suspended matter [səˌspendɪd 'mætə] Schwebstoff(e)
shingle ['ʃɪŋgl] Schotter
perforated ['pɜːfəreɪtɪd] durchlöchert
compressed air [kəm'prest eə] Druckluft
organism ['ɔːgənɪzəm] Organismus
purify ['pjʊərɪfaɪ] reinigen
beneficial [ˌbenɪ'fɪʃl] nützlich, vorteilhaft
prevention [prɪ'venʃn] Verhütung
disease [dɪ'ziːz] Krankheit
chalk well ['tʃɔːk wel] Kalkbrunnen
elsewhere [ˌels'weə] anderswo
chlorine ['klɔːriːn] Chlor
(trunk) mains ['trʌŋk meɪnz] Haupt(wasser)rohr(netz)
en route [ˌɒn 'ruːt] unterwegs
unobtrusively [ˌʌnəb'truːsɪvli] unaufdringlich, unauffällig
supplement ['sʌplɪment] ergänzen, ersetzen
gravity ['grævəti] Schwerkraft
consumer [kən'sjuːmə] Verbraucher/in, Konsument/in
household ['haʊshəʊld] Haushalt, Haushalts-
sample ['sɑːmpl] Muster, Probe
be conscious of sth [bi 'kɒnʃəs əv] sich einer Sache bewusst sein

Page 108

sorting technique ['sɔːtɪŋ tekniːk] Sortiertechnik, Scheidetechnik
sequence ['siːkwəns] Reihenfolge
recyclables [ˌriː'saɪkləblz] wiederverwertbare Stoffe, Wertstoffe
electromagnetic [ɪˌlektrəʊmæg'netɪk] elektromagnetisch
conveyor belt [kən'veɪə belt] Förderband
hydrocyclone [ˌhaɪdrəʊ'saɪkləʊn] Hydrozyklon
pellet ['pelɪt] Kügelchen, kleine Kugel
inclined [ɪn'klaɪnd] geneigt, schräg
spin [spɪn] spinnen, drehen
flake [fleɪk] Flocke
strip [strɪp] Streifen
grind [graɪnd] mahlen
density ['densəti] Dichte

Page 109

waste incineration [ˌweɪst ɪnˌsɪnə'reɪʃn] Müllverbrennung
municipal [mjuː'nɪsɪpl] Siedlungs-
costly ['kɒstli] kostspielig
exhaust gas [ɪg'zɔːst gæs] Abgas
ferromagnetic [ˌferəʊ mæg'netɪk] eisenmagnetisch
megawatt ['megəwɒt] Megawatt
trash [træʃ] Abfall, Müll
grate [greɪt] (Feuer-)Rost
melting point ['meltɪŋ pɔɪnt] Schmelzpunkt
(non-)combustible [ˌnɒn kəm'bʌstəbl] (nicht) brennbar(er Stoff)
boiler aggregate [ˌbɔɪlə 'ægrɪgət] Rückstände in der Brennkammer
sift [sɪft] (durch)sieben
scrubber ['skrʌbə] Teerscheider, Rieselturm
electrostatic precipitator [ɪlektrəʊˌstætɪk prɪ'sɪpɪteɪtə] elektrostatischer Ausfällapparat
particle ['pɑːtɪkl] Teilchen
electrostatically [ɪˌlektrəʊ'stætɪkli] elektrostatisch
attach to [ə'tætʃ tə] haften an, sich heften an
oppositely ['ɒpəzɪtli] gegensätzlich, entgegengesetzt
mechanically [mɪ'kænɪkli] mechanisch
fragment ['frægmənt] Fragment, Teilchen
condense [kən'dens] kondensieren
flue-gas ['flue gæs] Rauchgas
lime [laɪm] Kalk
acid gas [ˌæsɪd 'gæs] säurehaltiges Gas
hammermill shredder ['hæməmɪl ʃredə] Hammerwerkshredder
release [rɪ'liːs] freilassen, entlassen
magnetic separator [mægˌnetɪk 'sepəreɪtə] magnetische Trennstufe
Fahrenheit ['færənhaɪt] Fahrenheit
ignite [ɪg'naɪt] (sich) entzünden

Page 110

recyclability [ˌriːˌsaɪklə'bɪləti] Wiederverwertbarkeit
polymer ['pɒlɪmə] Polymer(e)
(non-)corrosive [ˌnɒn kə'rəʊsɪv] (nicht) korrodierend, (nicht) ätzend
molecule ['mɒlɪkjuːl] Molekül
thermoplastic [ˌθɜːməʊ'plæstɪk] thermoplastisch(er Stoff)
thermosetting [ˌθɜːməʊ'setɪŋ] durch Wärme härtbar
reheat [ˌriː'hiːt] wieder erwärmen
reform [ˌriː 'fɔːm] umformen
harden ['hɑːdn] härten, hart werden
reshape [ˌriː'ʃeɪp] umgestalten, umformen
homogeneous [ˌhɒmə'dʒiːniəs] homogen, gleichartig
desirable [dɪ'zaɪərəbl] wünschenswert
omnipresent [ˌɒmnɪ'preznt] allgegenwärtig
scientific [ˌsaɪən'tɪfɪk] (natur)wissenschaftlich
molecular [mə'lekjələ] molekular

Page 111

attractive force [əˌtræktɪv 'fɔːs] Anziehungskraft

slide along [slaɪd ə'lɒŋ] entlanggleiten
deterioration [dɪˌtɪəriə'reɪʃn] Verschlechterung
chemical ['kemɪkl] chemisch
subjected to [səb'dʒektɪd tə] unterworfen, ausgesetzt
heavily cross-linked [ˌhevili ˌkrɒs'lɪŋkt] stark miteinander verschränkt
rigid ['rɪdʒɪd] starr
bond [bɒnd] Band, Bindung
alter ['ɔːltə] (ver)ändern
recycler [ˌriː'saɪklə] Wiederverwerter
distant ['dɪstənt] weit (entfernt), fern
decomposition [ˌdiːkɒmpə'zɪʃn] Zersetzung, Zerfall, Verwesung
oversimplify [ˌəʊvə'sɪmplɪfaɪ] stark vereinfachen
strictly speaking ['strɪktli spiːkɪŋ] genau genommen

Page 112
switchboard ['swɪtʃbɔːd] Satzbautafel
throw-away society [ˌθrəʊəweɪ sə'saɪəti] Wegwerfgesellschaft

Page 113
dissolve [dɪ'zɒlv] (sich) auflösen
inferno [ɪn'fɜːnəʊ] Inferno
thoroughly ['θʌrəli] gründlich
devise [dɪ'vaɪz] ausdenken, entworfen
decompose [ˌdiːkəm'pəʊz] zerlegen, spalten, (sich) zersetzen
kiln [kɪln] Brennofen
char [tʃɑː] Holzkohle, verkohlte Masse
substance ['sʌbstəns] Substanz, Stoff
capture ['kæptʃə] auffangen
emission [ɪ'mɪʃn] Emission, Abgabe, Ausstoß
dramatically [drə'mætɪkli] dramatisch
highlight ['haɪlaɪt] hervorheben
dioxin [daɪ'ɒksɪn] Dioxin
(aromatic) hydrocarbon [ˌærə'mætɪk ˌhaɪdrə'kɑːbən] (aromatischer) Kohlenwasserstoff
groundwater ['graʊndwɔːtə] Grundwasser
estimate ['estɪmeɪt] schätzen
previous ['priːviəs] vorhergehend, bisherige/r/s
failure ['feɪljə] Misserfolg, Fehlschlag
disposer [dɪ'spəʊzə] Entsorger
sufficiently [sə'fɪʃntli] ausreichend, genügend, genug
currently ['kʌrəntli] momentan
pollutant [pə'luːtənt] Schadstoff
license ['laɪsns] genehmigen
toughest ['tʌfɪst] schärfst, härtest
refinement [rɪ'faɪnmənt] Verbesserung
nitrogen oxide [ˌnaɪtrədʒən 'ɒksaɪd] Stickoxid
sulphur dioxide [ˌsʌlfə daɪ'ɒksaɪd] Schwefeldioxid
incinerator [ɪn'sɪnəreɪtə] Verbrennungsanlage
national grid [ˌnæʃnəl 'grɪd] Überland-(leitungs)netz, Energieversorgungsnetz
operational [ˌɒpə'reɪʃənl] einsatzfähig, in Betrieb

Page 114
claim [kleɪm] behaupten, beanspruchen
banish ['bænɪʃ] (ver)bannen , vertreiben
discarded [dɪs'kɑːdɪd] ausrangiert
recyclable [ˌriː'saɪkləbl] wiederverwertbar, recycelbar
pile up [ˌpaɪl 'ʌp] (sich) anhäufen
researcher [rɪ'sɜːtʃə] Forscher/in
supercritical [ˌsuːpə'krɪtɪkl] superkritisch
merge [mɜːdʒ] (sich) verschmelzen, sich verbinden
sodium hydroxide [ˌsəʊdiəm haɪ'drɒksaɪd] Natriumhydroxyd, Ätznatron
organic [ɔː'gænɪk] organisch
reuse [ˌriː'juːz] wieder verwenden
molar ['məʊlə] molar, Mol(ar)-
compound ['kɒmpaʊnd] Verbindung
oily ['ɔɪli] ölig, Öl-
mixture ['mɪkstʃə] Mischung
textiles ['tekstaɪl] Textilien
recover [rɪ'kʌvə] wiedergewinnen, regenerieren
vulcanisation [ˌvʌlkənaɪ'zeɪʃn] Vulkanisierung
crosslink ['krɒslɪŋk] Querverbindung
hydrogen sulphide [ˌhaɪdrədʒən 'sʌlfaɪd] Wasserstoffsulfid
zinc oxide [zɪŋk 'ɒksaɪd] Zinkoxid
zinc sulphide [zɪŋk 'sʌlfaɪd] Zinksulfid
devulcanise [dɪ'vʌlkənaɪz] entvulkanisieren

Page 115
occur [ə'kɜː] vorkommen, enthalten sein
illegal [ɪ'liːgl] illegal, gesetzwidrig, verboten
graveyard ['greɪvjɑːd] Friedhof
alloy ['ælɔɪ] Legierung
tough [tʌf] widerstandsfähig, zäh, hart
mistake [mɪ'steɪk] missverstehen, verwechseln, irrtümlich halten für
abundant [ə'bʌndənt] reichlich (vorhanden)
shortage ['ʃɔːtɪdʒ] Knappheit, Mangel
agent ['eɪdʒənt] Mittel, Wirkstoff
purity ['pjʊərəti] Reinheit
elasticity [ˌiːlæ'stɪsəti] Elastizität, Biegsamkeit
improperly [ɪm'prɒpəli] falsch, unzulässig
inflexible [ɪn'fleksəbl] unflexibel, unbeweglich, starr
malleable ['mæliəbl] verformbar
elastic [ɪ'læstɪk] elastisch
dense [dens] dicht
prosperity [prɒ'sperəti] Prosperität, Wohlstand
solid ['sɒlɪd] fest
multiple ['mʌltɪpl] (das) Mehrfache, (das) Vielfache
transportation [ˌtrænspɔː'teɪʃn] Beförderung, Transport(wesen)

Page 116
mind mapping ['maɪnd mæpɪŋ] Vortragsnotizen
technician [tek'nɪʃn] Techniker/in
keyword ['kiːwɜːd] Schlüsselwort
overhead projector [ˌəʊvəhed prə'dʒektə] Overheadprojektor, Tageslichtprojektor
guidelines ['gaɪdlaɪnz] Richtlinien
blank [blæŋk] leer
frame of mind [ˌfreɪm əv 'maɪnd] Gemütsverfassung
inability [ˌɪnə'bɪləti] Unfähigkeit
image ['ɪmɪdʒ] Image, Bild
differentiate [ˌdɪfə'renʃieɪt] (sich) unterscheiden, differenzieren
condense [kən'dens] kondensieren, komprimieren
imaginative [ɪ'mædʒɪnətɪv] phantasievoll, einfallsreich
edit ['edɪt] redigieren, bearbeiten
regroup [ˌriː'gruːp] umgruppieren

Module 6

Page 117
ray [reɪ] Strahl
vibrating ['vaɪbreɪtɪŋ] vibrierend
gamma ray [ˌgæmə 'reɪ] Gammastrahl
accelerate [ək'seləreɪt] beschleunigen
pulse [pʌls] Puls
graphically speaking ['græfɪkli spiːkɪŋ] bildlich ausgedrückt
occur [ə'kɜː] geschehen, vorkommen
angle ['æŋgl] Winkel
broadcast ['brɔːdkɑːst] senden, übertragen
medium ['miːdiəm] Medium
substance ['sʌbstəns] Substanz
reversal [rɪ'vɜːsl] Umkehr(ung)
bounce off [ˌbaʊns 'ɒf] abprallen
transmission [træns'mɪʃn] Übertragung, Sendung
amplification [ˌæmplɪfɪ'keɪʃn] Verstärkung
at intervals [ət 'ɪntəvlz] dann und wann, ab und zu
receiver [rɪ'siːvə] Empfänger
referred to [rɪ'fɜːd tə] bezeichnet
device [dɪ'vaɪs] Vorrichtung, Gerät
beam [biːm] (Licht-)Strahl
distortion [dɪ'stɔːʃn] Verzerrung

Page 118
composed of [kəm'pəʊzd əv] zusammengesetzt aus
sub-system ['sʌbsɪstəm] Untersystem
perform [pə'fɔːm] durchführen, ausführen
encode [ɪn'kəʊd] verschlüsseln
decode [ˌdiː'kəʊd] entschlüsseln
retrieve [rɪ'triːv] wiederherstellen, wiedererlangen
brain [breɪn] Gehirn
transducer [træns'djuːsə] Umsetzer, Umwandler
earpiece ['ɪəpiːs] Hörmuschel
electrical current [ɪˌlektrɪkl 'kʌrənt] elektrischer Strom
modulation [ˌmɒdju'leɪʃn] Modulation
modulator ['mɒdjuleɪtə] Modulator
extract [ɪk'strækt] herausholen, gewinnen
coaxial cable [kəʊˌæksiəl 'keɪbl] Koaxialkabel
demodulate [ˌdiː'mɒdjuleɪt] demodulieren
amplifier ['æmplɪfaɪə] Verstärker
boost [buːst] erhöhen, verstärken
represent as [ˌreprɪ'zent əz] darstellen als
triangle ['traɪæŋgl] Dreieck
attenuation [əˌtenju'eɪʃn] Dämpfung, Abschwächung
random ['rændəm] Zufalls-
classify ['klæsɪfaɪ] klassifizieren, einteilen
artificial [ˌɑːtɪ'fɪʃl] künstlich
cosmic ['kɒzmɪk] kosmisch
atmospheric [ˌætməs'ferɪk] atmosphärisch
attenuating [ə'tenjueɪtɪŋ] dämpfend, abschwächend
amidst [ə'mɪdst] inmitten
repeater [rɪ'piːtə] (Leitungs-)Verstärker
regenerator [rɪ'dʒenəreɪtə] Regenerator

amplify ['æmplɪfaɪ] verstärken

Page 119

switchboard ['swɪtʃbɔ:d] Satzbautafel
determine [dɪ'tɜ:mɪn] beschließen, bestimmen
enable [ɪ'neɪbl] ermöglichen, befähigen
intend [ɪn'tend] beabsichtigen
permit [pə'mɪt] erlauben, zulassen
pretend [prɪ'tend] vorgeben, so tun als ob
deny [dɪ'naɪ] abstreiten, verneinen
postpone [pə'spəʊn] verschieben

Page 120

deteriorate [dɪ'tɪəriəreɪt] (sich) verschlechtern
proximity [prɒk'sɪməti] Nähe
precede [prɪ'si:d] vorausgehen, stehen vor
accommodation [ə,kɒmə'deɪʃn] Unterkunft, Zimmer, Wohnung, Wohnraum
permission [pə'mɪʃn] Erlaubnis, Einverständnis, Genehmigung

Page 121

noise-suppression [,nɔɪz sə'preʃn] Lärmschutz
power socket ['paʊə sɒkɪt] Steckdose (für Strom)
cherish ['tʃerɪʃ] schätzen, festhalten an, hegen
chat [tʃæt] Unterhaltung, Plauderei
airplane ['eəpleɪn] Flugzeug
commuter [kə'mju:tə] Pendler/in
cabin ['kæbɪn] Kabine, Passagierraum
operate in ['ɒpəreɪt ɪn] operieren in, eingesetzt werden in

Page 122

link [lɪŋk] Verbindung
wireless ['waɪələs] drahtlos, Funk-
thoroughly ['θʌrəli] gründlich
handset ['hændset] Handapparat
relay ['ri:leɪ] Relais, übertragen
remote [rɪ'məʊt] (weit) entfernt
dial ['daɪəl] wählen
(vice) chairman [,vaɪs'tʃeəmən] (stellvertretender) Vorsitzender, (Vize-)Präsident
recent ['ri:snt] kürzlich, jüngst
constellation [,kɒnstə'leɪʃn] Gruppe, Konstellation
orbit ['ɔ:bɪt] Umlaufbahn
iridium [ɪ'rɪdiəm] Iridium
nucleus ['nju:kliəs] Kern
polar orbit [,pəʊlər 'ɔ:bɪt] Umlaufbahn über den Pol
nautical mile [,nɔ:tɪkl 'maɪl] Seemeile, Knoten
upside-down [,ʌpsaɪd 'daʊn] verkehrt herum, umgekehrt
corporate ['kɔ:pərət] der Körperschaft
stationary ['steɪʃənri] stehend, parkend

Page 123

serial number ['sɪəriəl nʌmbə] Seriennummer
smart [smɑ:t] intelligent, klug
service center ['sɜ:vɪs sentə] Service Center
billing information [,bɪlɪŋ ,ɪnfə'meɪʃn] Abrechnungsinformationen
authorise ['ɔ:θəraɪz] autorisieren
beam [bi:m] strahlen
via ['vaɪə] über, durch
disaster relief worker [dɪ,zɑ:stə rɪ'li:f wɜ:kə] Katastrophenhelfer/in
aboard [ə'bɔ:d] an Bord
estimated ['estɪmeɪtɪd] geschätzt
page someone [peɪdʒ] jdn ausrufen
magic ['mædʒɪk] Zauberei, Zauber
urban ['ɜ:bən] städtisch, Stadt-
invest [ɪn'vest] investieren
handheld communicator [,hændheld kə'mju:nɪkeɪtə] Handy
two-way data pager [,tu:weɪ 'deɪtə peɪdʒə] Datenübertrager und -empfänger
desktop PC ['desktɒp ,pi: 'si:] Bürocomputer
mainframe computer [,meɪnfreɪm kəm'pju:tə] Großrechner
portable ['pɔ:təbl] transportabel, tragbar
hence [hens] daher
tap into [tæp 'ɪntə] sich einschalten, anzapfen

Page 124

succinctly [sək'sɪŋktli] prägnant, in Kürze
suppress [sə'pres] unterdrücken
recreate [,ri:kri'eɪt] neu (er)schaffen
ignore [ɪg'nɔ:] ignorieren, nicht beachten
cord [kɔ:d] Schnur, Kabel
coaching ['kəʊtʃɪŋ] Anweisungen
protocol ['prəʊtəkɒl] Protokoll
mode [məʊd] Form, Modus
bulletin board ['bʊlətɪn bɔ:d] Anzeigetafel
electronical(ly) [,ɪlek'trɒnɪkli] elektronisch
intriguing [ɪn'tri:gɪŋ] spannend, faszinierend
venture ['ventʃə] (sich) wagen
minivan ['mɪnivæn] Kleinbus
makings ['meɪkɪŋz] Einrichtung
optional ['ɒpʃənl] wahlweise
cellular kit [,seljələ 'kɪt] Mobilfon
gather ['gæðə] sammeln
presumably [prɪ'zju:məbli] vermutlich
eliminate [ɪ'lɪmɪneɪt] eliminieren, beseitigen
solid ['sɒlɪd] fest
reassuring [,ri:ə'ʃʊərɪŋ] vertraut
screech [skri:tʃ] Schrillen
endure [ɪn'djʊə] aushalten, erfahren, erdulden

Page 125

fluke [flu:k] Glücksfall, Zufall
consistent(ly) [kən'sɪstəntli] einheitlich
file [faɪl] Datei, Ordner
hitch [hɪtʃ] Störung, Hindernis
reject [rɪ'dʒekt] zurückweisen, abweisen
error-controlling [,erə kən'trəʊlɪŋ] Fehlerkontrolle
surprising(ly) [sə'praɪzɪŋli] überraschend(erweise)
rate [reɪt] Rate
driveway ['draɪvweɪ] Auffahrt
level geography [,levl dʒi'ɒgrəfi] ebene Bodenbeschaffenheit
boast [bəʊst] verfügen über, aufzuweisen haben
spotty ['spɒti] ungleichmäßig, uneinheitlich
glitch [glɪtʃ] Ausfall, Panne
transform [træns'fɔ:m] verwandeln
informed [ɪn'fɔ:md] informiert, unterrichtet
(hi-fi) enthusiast [,haɪfaɪ ɪn'θju:ziæst] Hi-Fi-Anhänger / Enthusiast/in, begeisterter Hi-Fi Hörer

Vocabulary: alphabetical list

A

ability *29* Fähigkeit
aboard *123* an Bord
absorption *85* Aufnahme
abundant *115* reichlich (vorhanden)
accelerate *117* beschleunigen
acceleration *30* Beschleunigung
(un)acceptable *53* (un)annehmbar, (in)akzeptabel
access *29* Zugang, Zugriff
accessories *15* Zubehör(teile)
accidental *51* zufällig
accommodate *36* unterbringen, beherbergen
accommodation *120* Unterkunft, Zimmer, Wohnung, Wohnraum
accompanied by *103* begleitet von
accomplish *95* schaffen, erreichen
according to *33* entsprechend, nach
accordingly *33* (dem)entsprechend
achieve *49* erreichen, erlangen
achievement *101* Leistung, Errungenschaft
acid gas *109* säurehaltiges Gas
additional *81* zusätzlich
addressee *38* Empfänger/in, Adressat/in
adequate *85* ausreichend, adäquat
adhesive tape *98* Klebeband, Heftpflaster
adjacent *62* angrenzend, benachbart
adjoining *62* anliegend, benachbart
adjust *51* anpassen, einstellen
adjustable *101* verstellbar
adjustment *63* Einstellung, Anpassung
advance *45* Fortschritt
advanced *29* fortschrittlich
advent *101* Ankunft, Aufkommen
adverse *85* ungünstig, nachteilig, negativ
advertising feature *16* Werbeargument
adviser *73* Berater/in
aerodynamic *102* aerodynamisch
affect *19* beeinflussen, sich auswirken auf
agent *115; 87* Mittel, Wirkstoff; Handelnde/r
airplane *102* Flugzeug
align *94* ausrichten
allocate *62* zuweisen, zuteilen
alloy *84* Legierung
alter *111* (ver)ändern
alteration *16* Änderung
alternate *100* abwechselnd
alternator *90* Wechselstromgenerator
aluminium *17* Aluminium
ambient *80* umgebend
ambitious *36* ehrgeizig
amidst *118* inmitten
amphibious *100* amphibisch, Amphibien-
amplification *72* Verstärkung
amplifier *118* Verstärker
amplify *72* verstärken
angle *117* Winkel
annual(ly) *22* jährlich
anticipate *33* voraussehen
anticipation *101* Erwartung
anti-freeze *82* Gefrier(mittel), Frostschutz(mittel)
apparently *48* offensichtlich, anscheinend
appearance *97* Äußeres, Aussehen
appliance *12* Gerät
appoint *29* einstellen, ernennen, berufen
apprenticeship *59* Lehre, Ausbildung
appropriate *28* geeignet, passend
approve *17* gutheißen, billigen, genehmigen
arch *101* Bogen
arch bridge *101* Bogenbrücke
area of coverage *62* Reichweite
arguably *73* darüber lässt sich streiten
armature coil *95* Ankerwicklung
arrangement *47* Absprache, Vereinbarung
artificial *118* künstlich
assemble *14* zusammensetzen, zusammenbauen
assembler *87* Monteur
assembly *84* Gruppe
(sub)assembly line *73* (Neben-)Fließband
associate *12* assoziieren, verbinden
association *37* Verband, Gesellschaft
assume *55* annehmen, übernehmen, glauben
assure *24* versichern, zusichern
at a rate of *107* mit einer Geschwindigkeit von
at intervals *117* dann und wann, ab und zu
atmospheric *118* atmosphärisch
(sub)atmospheric pressure 82 (Unter-) Luftdruck
attach to *67; 109* befestigen an, verbinden mit; haften an, sich heften an
attenuating *118* dämpfend, abschwächend
attenuation *118* Dämpfung, Abschwächung
attitude *42* Einstellung, Haltung
attractive force *111* Anziehungskraft
authorise *123* autorisieren
authority *42* Verwaltung, Behörde, Amt
automotive *45* Auto-
autotechnology *45* Kfz-Technik
autoworker *74* Kfz-Arbeiter/in
auxiliary *79* zusätzlich
available in *17* erhältlich in, verfügbar in
axis *33* Achse
axle *50* Achse

B

bacteria *18* Bakterien
badly fitting *17* schlecht schließend
banish *114* (ver)bannen , vertreiben
bare *50* nackt, bloß
barrier *84* Schranke
be conscious of sth *107* sich einer Sache bewusst sein
be proportional to *69* proportional sein zu, im Verhältnis stehen zu
beam *98; 123* (Licht-)Strahl; strahlen
bearings *67* Lager
beneficial *107* nützlich, vorteilhaft
benefit *16* Nutzen, Vorteil
bid *35* (An-)Gebot
bid for *28* ein (An-)Gebot machen
billing information *123* Abrechnungsinformationen
blade *83* Blatt, Schaufel
blank *18; 116* Lücke; leer
boast *125* verfügen über, aufzuweisen haben
body shell *50* Karosserie
boiler *17* Heißwasserspeicher, (Heiz-) Kessel
boiler aggregate *109* Rückstände in der Brennkammer
bolt *31* Bolzen, Schraube
bond *111* Band, Bindung
boost *118* erhöhen, verstärken
borehole *107* Bohrloch
boric acid *84* Borsäure
borne by *101* getragen durch
bounce *104* (ab)prallen
bounce off *117* abprallen
box-section *98* mit quadratischem Querschnitt
brain *118* Gehirn
brake lining *73* Bremsbelag
braking fluid *18* Bremsflüssigkeit
breakthrough *17* Durchbruch
brick *17* Ziegelstein, Backstein
brief *90* kurz
briefly *10* kurz
broadcast *117* senden, übertragen
brush *90* (Kontakt-)Bürste, Stromabnehmer
building code *36* Bauvorschriften
bulletin board *124* Anzeigetafel
bumper *49* Stoßstange
burst *17* (Wasser-)Rohrbruch

C

cabin *121* Kabine, Passagierraum
(cold) calculation *42* (kühle) Berechnung
callout service *17* Notdienst
calorific value *50* Heizwert
camshaft *66* Nockenwelle
candy *44* Bonbon, Praline, Süßigkeiten
cantilever *101* aufhängen
cantilever bridge *101* Auslegerbrücke
capacity *85; 104* Kapazität, Leistung, Fähigkeit; Kapazität, Fassungsvermögen, Platz
capture *113* auffangen
car interior *17* Wageninneres
carbon dioxide *77* Kohlendioxid
carbon monoxide *18* Kohlenmonoxid
(torch) casing *93* (Lampen-)Gehäuse
cast *98* gießen
casting *100* Guss, Gießen
casual *53* leger, sportlich
category *69* Kategorie, Klasse
cavity *17* Hohlraum
celebrate *35* feiern
cell site *63* Zelle/Zellenstandort
cellular *13* Zelle(n)-
cellular kit *124* Mobilfon
cellular telephony *63* Mobilfunk
cement *17* Zement
centrally *62* zentral
CFC *23* FCKW
(vice) chairman *122* (stellvertretender) Vorsitzender, (Vize-)Präsident
chairperson *86* Vorsitzende/r, Moderator/in
chalk well *107* Kalkbrunnen
chamber *82* Kammer

char *113* Holzkohle, verkohlte Masse
character trait *64* Charakterzug
chart *32* Diagramm, Tabelle
chat *121* Unterhaltung, Plauderei
chemical *111* chemisch
cherish *121* schätzen, festhalten an, hegen
chlorine *107* Chlor
circuit *70* Schaltung
circulating pump *84* Umlaufspumpe
City and Guilds *60* Eine Prüfungskammer, die eigene Industrie/Handwerksprüfungen durchführt.
(zirconium-)clad *84* (mit Zirkonium) verkleidet
claim *51* beanspruchen, behaupten
clamp together *101* festklemmen
clarify *50* (auf)klären
classify *69* klassifizieren, einteilen
client *18* Kunde, Kundin
(anti)clockwise *68* im/gegen den Uhrzeigersinn
coaching *124* Anweisungen
coarse *107* grob
coaxial cable *118* Koaxialkabel
coil *78* Windung, Spule
coil meter *88* Spannungsmesser
collapse *102* Zusammenbruch, Einsturz
collector *77* Sammler, Kollektor
column *64* Spalte, Kolumne
(non-)combustible *109* (nicht) brennbar (-er Stoff)
combustion *29* Verbrennung
come out of the blue *55* aus heiterem Himmel kommen
commercial(ly) *51* kommerziell, wirtschaftlich
commission *36* Kommission, Auftrag
commitment *25* Verpflichtung, Engagement
commonly *14* häufig, gewöhnlich, üblicherweise
communicator *64* Unterhalter/in
commutator *95* Kollektor, Stromwender
commuter *121* Pendler/in
component *50* Bauteil, Bestandteil
composed of *118* zusammengesetzt aus
compound *114* Verbindung
compress *98* komprimieren, zusammenpressen
compressed air *107* Druckluft
compression *98* Kompression, Druck
compressive *98* Press-, Druck-
comprise *29* umfassen, bestehen aus
concentrate on *75* sich konzentrieren auf
concrete *28* konkret
condensate *85* Kondensat
condense *109; 116* kondensieren; komprimieren
condenser *85* Kondensator
conduct *74* durchführen
conducted by *42* durchgeführt von
conductor *90* Konduktor
confer *63* sich abstimmen
confirm *48* bestätigen
conform to *17* entsprechen
consequent *93* (dem)entsprechend
consequently *63* dementsprechend, infolgedessen
conservatory *16* Gewächshaus
conserve *17* konservieren, erhalten
considerable (ably) *50* beträchtlich
consideration *87* Erwägung, Überlegung
consistent(ly) *125* einheitlich
constellation *122* Gruppe, Konstellation
constitute *84* darstellen
constraint *37* Einschränkung, Beschränkung
construction *36* Bau, Konstruktion
consult *66* konsultieren, zu Rate ziehen, nachschlagen in
consultant *37* Berater/in, Gutachter/in
consumer *51* Konsum-, Verbraucher/in, Konsument/in
consumption *16* Konsum, Verbrauch
contain *18* enthalten
contamination *49* Verseuchung, Kontaminierung, Verunreinigung
content *102* Gehalt, Inhalt
contract *28; 104* Vertrag, Kontrakt; zusammenziehen
contractor *100* Unternehmer (mit Werkvertrag)
contribution *41* Beitrag
convenient *39* passend, günstig, bequem
conventional *31* herkömmlich
conversely *63* umgekehrt
convert *29* umwandeln, umbauen
conveyor belt *108* Förderband
cooperate *29* zusammenarbeiten
cooperative *33* kooperativ
copper *51* Kupfer
cord *124* Schnur, Kabel
cord grip *92* Kabelhalter
core *97* Kern
corporate *122* der Körperschaft
correspond to *77* entsprechen
(non-)corrosive *110* (nicht) korrodierend, (nicht) ätzend
cosmic *118* kosmisch
costly *109* kostspielig
counter-balanced *101* (durch Gegengewicht) ausgeglichen
counter-clockwise *91* gegen den Uhrzeigersinn, links herum
crack *70* Riss, Spalt(e)
craft *100* Fahrzeug(e), Schiff(e)
craftsman *17* (Kunst-)Handwerker/in
cranage *104* Krangebühr
crank *67* Kurbel
crankshaft *66* Kurbelwelle
creativity *43* Kreativität
critical *76* kritisch
criticism *53* Kritik
crosslink *114* Querverbindung
cross-section *98* Querschnitt
crumble *98* zerfallen, zerbröseln
cubic-(foot) *77* Kubik(fuß) (1 cu foot = 0.028 m^3)
currently *113* momentan
cylindrical *95* zylindrisch

D

damping element *88* Dämmelement
dashboard *50* Armaturenbrett
dealership *49* Vertretung
decision-making *12* Treffen von Entscheidungen
deck segment *100* Plattformsegment
decode *118* entschlüsseln
decompose *113* zerlegen, spalten, (sich) zersetzen
decomposition *111* Zersetzung, Zerfall, Verwesung
decorate *17* tapezieren, renovieren
defective *70* fehlerhaft
(in)definite *43* (un)bestimmt
deflect *98* ablenken, umbiegen, abweichen
deflection movement *104* Ablenkungsbewegung
demodulate *118* demodulieren
demoralize *74* demoralisieren, entmutigen
dense *115* dicht
density *108* Dichte
deny *119* abstreiten, verneinen
depict *106* darstellen
derive from *92* ableiten von
derogatory *76* abfällig, abträglich
desirable *110* wünschenswert
desktop PC *123* Bürocomputer
despite *79* trotz
destruction *102* Zerstörung, Vernichtung
detect *18* entdecken
detector *62* Detektor
deteriorate *120* (sich) verschlechtern
deterioration *111* Verschlechterung
determine *63* bestimmen, feststellen, beschließen
determined by *42* bestimmt von
device *67* Vorrichtung, Gerät, Regler
devise *113* ausdenken, entworfen
devulcanise *114* entvulkanisieren
diagnostic *74* diagnostisch
dial *122* wählen
diameter *66* Durchmesser
die *66* (Draht-)Ziehstahl
differ *23* sich unterscheiden
differential thermostat *83* Differentialthermostat
differentiate *116* (sich) unterscheiden, differenzieren
digitize *70* digitalisieren
dioxin *113* Dioxin
directly *69* direkt
disassembly *94* Demontage, Zerlegung
disaster relief worker *123* Katastrophenhelfer/in
discarded *114* ausrangiert
discrete *63* getrennt, einzeln
disease *107* Krankheit
dismantle *49* abwracken, zerlegen, demontieren
disorder *105* Störung, Unordnung
displace *77* ersetzen
disposal *85* Beseitigung, Entsorgung
dispose of *93* beseitigen, entsorgen
disposer *113* Entsorger
disregard *64* ignorieren, nicht beachten
disrupt *104* unterbrechen
dissipated *85* zerstreut, aufgelöst
dissolve *84* (sich) auflösen
distant *111* weit (entfernt), fern
distinct *88* deutlich, klar erkennbar
distinction *41* Unterschied, Unterscheidung
distinguish *41* unterscheiden
distortion *117* Verzerrung
distribute *51* verteilen
division *45* Abteilung
DIY enthusiast *17* Heimwerker
domed *104* mit einer Kuppel versehen, kuppelförmig
domestic *12* für das Inland (gedacht), Inlands-
dominate *36* beherrschen
double glazed *17* mit Isolierverglasung
downwind *104* in Windrichtung

draft *105* Entwurf
drain *49* absaugen, ablaufen lassen
drain back *78* zurücksaugen, zurückfließen
drain down *107* absaugen, abfließen
drain pipe *79* Abflussrohr, Abzugsröhre
dramatically *113* dramatisch
drastically *33* drastisch
draught *17* Zugluft
draught stripping *17* Isolierband (gegen Luftzug)
drawing *66* Ziehen, Walzen
drive *85* antreiben
drive transmission *29* Triebwelle
driveway *125* Auffahrt
drought *107* Trockenzeit, Dürre(periode)
durable *28* dauerhaft, haltbar
dustbin *16* Mülltonne

E

earpiece *118* Hörmuschel
echo *36* wieder aufnehmen
ecological *77* ökologisch
economy *53* Wirtschaft
edit *116* redigieren, bearbeiten
effective(ly) *17* effektiv, wirksam, wirkungsvoll
efficiency *10* Effizienz, Leistungsfähigkeit
efficient *17* effizient, (leistungs)fähig, tüchtig
elastic *115* elastisch
elasticity *115* Elastizität, Biegsamkeit
electric *11* elektrisch
electrical(ly) *95* elektrisch
electrical current *118* elektrischer Strom
electromagnetic *108* elektromagnetisch
electromotive force (e.m.f) 90 elektromotorische Kraft
electronical(ly) *124* elektronisch
electrostatic precipitator 109 elektrostatischer Ausfällapparat
electrostatically *109* elektrostatisch
eliminate *124* eliminieren, beseitigen
elsewhere *107* anderswo
emancipation *42* Emanzipation
emission *16* Emission, Abgabe, Ausstoß
emphasis *10* Betonung, Nachdruck
emphasize *44* betonen, hervorheben
employ *59* anstellen, beschäftigen, einstellen
en route *107* unterwegs
enable *68* ermöglichen, befähigen
Encl. *24* Anlage(n)
enclose *7* beilegen, beifügen
encode *118* verschlüsseln
endure *124* aushalten, erfahren, erdulden
energy source *16* Energiequelle
Eng. *24* Ingenieur
engineering know-how *28* technischer Sachverstand
enrich *84* anreichern
ensure *17* sicherstellen, gewährleisten
(hi-fi) enthusiast *125* Hi-Fi-Anhänger / Enthusiast/in, begeisterter Hi-Fi Hörer
entire *74* ganz
environmental(ly) *45* Umwelt-
equal(ly) *36* gleich(ermaßen)
equip *11* ausrüsten, ausstatten
equipment *19* Ausrüstung, Ausstattung, Geräte
equivalent *56* entsprechend
erect *101* (er)bauen, errichten
error-controlling *125* Fehlerkontrolle
espagnolette handle *17* Drehriegel
essential *44* (absolut) notwendig, wichtig, wesentlich, (unbedingt) erforderlich
estimate *16; 113* Schätzung, (Kosten-)Anschlag; schätzen
estimated *123* geschätzt
etc *9* usw.
etiquette *44* Etikette
evaluate *94* (ein)schätzen, bewerten
eventually *62* schließlich, zum Schluss
evidence *76* Beweismaterial, Beweis(e)
evolve *97* entwickeln, erzeugen
exceed *78* übersteigen
excess *16* übermäßig, im Übermaß
excessive *104* übermäßig, übertrieben
(heat) exchanger *78* Wärmetauscher
exert *91* ausüben, anwenden
exhaust gas *109* Abgas
exhibit *45* ausstellen
exhibition *45* Ausstellung
expand *11* erweitern, expandieren
expansion *11* Ausdehnung, Erweiterung, Expansion
expectation *74* Erwartung
expenditure *16* Ausgabe
expertise *11* Sachkenntnis
export *15* ausführen, exportieren
expose *92* freilegen
extend *29* erweitern, ausdehnen, vergrößern
extension number *20* Nebenanschluss, Direktwahl
extensive *28* umfassend
extract *30; 118* Auszug; herausholen, gewinnen
extreme *97* äußerst, extrem, außerordentlich

F

facade *36* Fassade
facility *10* Einrichtung
(south)-facing *36* nach (Süden) liegend
Fahrenheit *101* Fahrenheit
failure *113* Misserfolg, Fehlschlag
fall *47* Herbst
fan *28* Gebläse, Ventilator
fan out *104* sich (fächerförmig) ausbreiten, sich erstrecken
fascinating *62* faszinierend, hochinteressant
fast-acting *29* schnell arbeitend
fasten *31* befestigen, anziehen
fastened to *67* befestigt an
feather *97* Feder
feature *28* Merkmal, Kennzeichen
ferromagnetic *109* eisenmagnetisch
(non-)ferrous *50* (Nicht-)Eisen-
file *125* Datei, Ordner
filtration *107* Filtern, Filtrierung
finalise *105* endgültig festlegen, abschließen
financially *23* finanziell, wirtschaftlich
(oil-)fired *16* Heizöl-
fission *84* Spaltung
fitter *17* Monteur/in, (Maschinen-)Schlosser/in, Installateur/in
flake *108* Flocke
flatten out *33* (sich) abflachen
fleet *100* Flotte
flex clamp *92* Litzenschelle
flexibly *68* flexibel
float *82* Schwimmer
floating debris *107* schwimmende Teile
flow-job *13* Fließfertigung
fluctuate *32* schwanken
fluctuation *32* Schwankung, Ausschlag
flue-gas *109* Rauchgas
fluent *60* fließend
fluid *49* Flüssigkeit
fluke *125* Glücksfall, Zufall
flywheel *66* Schwungrad
foam *17* Schaumstoff
foot treadle *66* Fußantrieb
for clarity *95* aus Gründen der Klarheit
forefront *73* Spitze
(in)formal *8* (in)offiziell, (in)formell
former *58* ehemalig, früher
formulate *52* formulieren, ausarbeiten
fossil *85* fossil
found *12* gründen, begründen
foundation *17; 102* Stiftung, Anstalt; Grundlage, Fundament
fraction *62* Bruchteil
fragile *31* zerbrechlich
fragment *109* Fragment, Teilchen
(steel-)framed *104* Stahlbau-
frame of mind *116* Gemütsverfassung
freeze-up *78* Einfrieren
frequency *62* Frequenz
frequency pair *63* Frequenzpaar
frequency synthesizer *63* Frequenzsynthesizer
friction *67* Reibung
frost-protecting *78* gegen Frost schützen
frustrated *57* frustriert
fulfil *23* erfüllen
functional *42* funktional, zweckmäßig
furthermore *90* ferner, überdies
fuse *92* Sicherung
fuseholder *93* Sicherungsfassung

G

gain *79* gewinnen
galloping *102* galoppierend
galvanized steel *102* verzinkter Stahl
gamma ray *117* Gammastrahl
gas-cushioned pressurizer 84 Druckausgleichsbehälter
gather *124* sammeln
gearbox *50* Getriebe
gently sloping *36* sanft abfallend
get at *33* hinauswollen auf
glazing *16* Verglasung
glitch *125* Ausfall, Panne
global warming *86* Erderwärmung, Erwärmung der Erdatmosphäre
go slack *104* schlaff werden, erschlaffen
governed by *36* verwaltet von, geregelt durch
gradual(ly) *17* allmählich, nach und nach
granule *72* Granulat
graphically speaking *117* bildlich ausgedrückt
grate *109* (Feuer-)Rost
graveyard *115* Friedhof
gravity *107* Schwerkraft
grease *92* Fett, (ein)fetten
grid *98* Gitter
grind *108* mahlen
grinding machine *14* Schleifmaschine
grip *67* (er)greifen
gripper *70* Greifer
groundwater *113* Grundwasser
group mentality *64* Gruppenmentalität
guidelines *116* Richtlinien
guy *104* Halteseil

H

hammermill shredder *109* Hammerwerkshredder
handheld communicator *123* Handy
handing-off *62* Übergabe-
handset *122* Handapparat
handshake *14* Händedruck
harden *110* härten, hart werden
hardware *17* Haushaltswaren, Eisenwaren
harmonious *64* harmonisch
hazard *82* Risiko, Gefahr
header *79* Rohrverbinder
headquarters *56* Hauptquartier, Zentrale
heat treat *14* warmbearbeiten, vergüten
heavily cross-linked *111* stark miteinander verschränkt
heavy-walled *84* dickwandig
hence *123* daher
hermetically sealed *84* hermetisch verschlossen
highlight *113* hervorheben
high-rise *104* Hochhaus-
hinge *36* Scharnier
hitch *125* Störung, Hindernis
hoist up *104* hochziehen
hollow *98* hohl
homogeneous *110* homogen, gleichartig
hopper *72* Trichter
hoseproofed *28* schlauchisoliert
household *107* Haushalt, Haushalts-
(aromatic) hydrocarbon *113* (aromatischer) Kohlenwasserstoff
hydrocyclone *108* Hydrozyklon
hydrogen sulphide *114* Wasserstoffsulfid

I

ignite *109* (sich) entzünden
ignore *124* ignorieren, nicht beachten
illegal *115* illegal, gesetzwidrig, verboten
illustrate *29* illustrieren, veranschaulichen
illustrated by *16* veranschaulicht durch
image *116* Image, Bild
imaginary *98* imaginär, gedacht
imaginative *116* phantasievoll, einfallsreich
immediate *37* sofortige(r,s), umgehend
immobile *64* unbeweglich
impact *85* Wirkung, Einfluss
impart *83* mitteilen, übertragen
imperfection *70* Unvollkommenheit, Mangel
import *15* einführen, importieren
impressed by *24* beeindruckt durch
impressive *29* beeindruckend
improperly *115* falsch, unzulässig
improvement *16* Verbesserung
impurity *107* Verunreinigung, Unreinheit
in contrast *51* im Gegensatz (dazu)
in such a manner *67* so
inability *116* Unfähigkeit
inadequate *102* nicht ausreichend, ungenügend
incinerator *113* Verbrennungsanlage
inclined *108* geneigt, schräg
including (incl.) *35* einschließlich, inbegriffen
inconvenience *26* Unbequemlichkeit
incorporate *79* enthalten
incorrect(ly) *18* falsch, unrichtig
index card *87* Karteikarte
indicate *12* anzeigen
induce *90* auslösen, induzieren
induction motor *29* Induktionsmotor
inferno *113* Inferno
inflexible *115* unflexibel, unbeweglich, starr
influential *41* einflussreich
informed *125* informiert, unterrichtet
ingenious *42* klug (erdacht), genial
in-house *51* innerbetrieblich
initial costs *22* Initialkosten, Gestehungskosten
initially *81* anfänglich, zuerst
initiative *50* Initiative
inlet *82* Einlass, Öffnung
innovation *11* Neuerung, Innovation
input *81* Eingabe
inquire about *61* sich erkundigen nach, fragen nach
insert *70* einfügen
inspect *70* inspizieren
installer *17* Installateur
institutional *42* institutionell
instruct *63* anweisen
instruction *63* Anweisung
instructor *33* Lehrer/in, Ausbilder/in
insulate *17* isolieren
insulation *17* Isolierung, Isolation
intact *93* intakt
integral part *36* fester Bestandteil
integrate *37* eingliedern, integrieren
intend *119* beabsichtigen
intensity *80* Intensität
intention *47* Absicht
interface *58* Schnittstelle
interfere *63* stören
interference *63* Störung
interior *49* Inneres, Innenausstattung
intermittent(ly) *68* mit Unterbrechungen
interpret *41* interpretieren
interval *87* Abstand, Pause
intriguing *124* spannend, faszinierend
inventor *82* Erfinder/in
invest *123* investieren
investment *11* Investition, Anlage
involved in *11* beteiligt an
involvement *37* Beteiligung
iridium *122* Iridium
irrelevant *14* unwesentlich, nebensächlich
irreplaceable *16* unersetzlich
isolate *18* isolieren
issue *11* Frage, Angelegenheit, Problem, Thema

J

jack *104* Hebevorrichtung
jaws *67* Backen, Maul
job-shop *13* Werkstattfertigung
joint *79; 104* Verbindungsrohr; gemeinsam
journal *37* Zeitschrift
jumbled-up *9* durcheinander (gewürfelt)

K

key locking *17* abschließbar
keyword *116* Schlüsselwort
kiln *113* Brennofen

L

labor *74* Arbeit, Arbeitskräfte, Arbeitskraft
laboratory *58* Labor(atorium)
land lines *56* Festnetz
landfill *50* geordnete Mülldeponie
last but not least *52* nicht zuletzt
later *14* später
lateral *96* seitlich
launch *11* einführen, starten, auf den Markt bringen
layout *40* Layout, Grundriss
leadership *55* Führungsqualität
leak *17* undichte Stelle, Leck
lean *12* schlank
legislation *45* Gesetzgebung, Gesetze
lengthen *67* verlängern
level geography *125* ebene Bodenbeschaffenheit
lever out *92* (her)aushebeln
liberate *42* befreien
license *113* genehmigen
light bulb *90* Glühbirne
lightness *97* Leichtigkeit
lime *109* Kalk
link *79; 122* verbinden; Verbindung
linking word *82* Bindewort
load *28* Ladung, Last, Fracht, Belastung
load-bearing pin *104* die Last tragender Bolzen
location *12* Standort, Lage
loft *17* (Dach-)Boden
look forward to *37* sich freuen auf
loop *77* Schleife
Ltd *16* mit beschränkter Haftung, Gesellschaft mit beschränkter Haftung, GmbH
lubricate *71* schmieren
lubrication *67* Schmierung

M

magic *123* Zauberei, Zauber
magnetic *88* magnetisch
magnetic separator *109* magnetische Trennstufe
magnetics *28* Wissenschaft vom Magnetismus
magnitude *90* Größe
mahogany *16* Mahagoni
mainframe computer *123* Großrechner
mainly *29* hauptsächlich
(trunk) mains *107* Haupt(wasser)rohr(netz)
maintain *28; 74* aufrechterhalten, bewahren; warten
maintenance *28* Wartung, Pflege
make redundant *75* entlassen, freisetzen
makings *124* Einrichtung
malleable *115* verformbar
management *8* Management, Geschäftsführung
manual *76; 94* mit der Hand; Handbuch
manufacturer *7* Hersteller/in, Produzent/in, Fabrikant
manufacturing *7* Herstellung, Produktion
march *73* Marsch
marital status *60* Familienstand
massively *29* massiv
masterpiece *42* Meisterstück, Meisterwerk
mathematical model *41* mathematisches Modell
mechanical engineer *45* Maschinen(bau)-techniker/in
mechanically *109* mechanisch
mechanism *66* Mechanismus
mechanization *67* Mechanisierung
medieval *67* mittelalterlich
medium *117* Medium
megawatt *109* Megawatt
melting *72* Schmelzen, Schmelz-
melting point *109* Schmelzpunkt
melt-pumping zone *72* Schmelzdruckbereich

merge *114* (sich) verschmelzen, sich verbinden
metallurgy *101* Metallurgie, Hüttenkunde
micron *74* Mikron, Mikrometer
microprocessor *70* Mikroprozessor
midspan *101* Mittelteil (einer Auslegerbrücke)
millisecond *63* Millisekunde
mind mapping *116* Vortragsnotizen
minimal *81* minimal
minimize *101* verkleinern, verringern
minivan *124* Kleinbus
mistake *115* missverstehen, verwechseln, irrtümlich halten für
mixture *114* Mischung
mock-up *36* Modell in natürlicher Größe
mode *124* Form, Modus
modifier *33* Bestimmungswort
modulation *118* Modulation
modulator *118* Modulator
molar *114* molar, Mol(ar)-
molecular *110* molekular
molecule *110* Molekül
monitor *63* überwachen
moulded plug *93* eingeschweisster Stecker
mount (on) *67* montieren (auf)
multi-(national) *45* multi(national)
multiple *115* (das) Mehrfache, (das) Vielfache
multipurpose *58* Vielzweck-
municipal *109* Siedlungs-

N

nappy *57* Windel
national grid *113* Überland(leitungs)netz, Energieversorgungsnetz
nautical mile *122* Seemeile, Knoten
negligible *31* geringfügig
neutron *84* Neutron
neutron absorbing boron *84* das Neutronen absorbierende Bor
nitrogen oxide *113* Stickoxid
noise-suppression *121* Lärmschutz
no-parking zone *27* Parkverbotszone
noxious *16* schädlich
nuclear energy *23* Kernenergie
nucleus *122* Kern
numeric *69* numerisch

O

objective *12* Ziel
observe *52* beachten
obstacle *42* Hindernis
obtain *16* erhalten, erzielen
obtainable *85* erhältlich, zu beziehen
occasional(ly) *14* gelegentlich
occur *98; 115* geschehen, vorkommen; enthalten sein
oily *114* ölig, Öl-
omission *94* Auslassen
omit *85* auslassen, weglassen
omnipresent *110* allgegenwärtig
on display *40* ausgestellt
open/closed-loop *78* mit offener/geschlossener Schleife
operate in *121* operieren in, eingesetzt werden in
operational *113* einsatzfähig, in Betrieb
operator *66* Maschinist/in
oppositely *109* gegensätzlich, entgegengesetzt
optical fibre *26* Glasfaser
optically *69* optisch
optimum *81* Optimum, optimal
optional *124* wahlweise
orbit *122* Umlaufbahn
organic *114* organisch
organism *107* Organismus
oscillating *68* oszillierend
outcry *16* Aufschrei
outer sheath *92* Ummantelung
outlet *82* Abfluss, Ausgang
output *29* Produktion(smenge), Ausstoß
outstanding *28* hervorragend
outward *96* (nach) außen
overhead *73* oben, darüber
overhead projector *116* Overheadprojektor, Tageslichtprojektor
overhead transparency *79* Tageslichtfolie
oversimplify *111* stark vereinfachen
overview *101* Überblick
owing to *26* wegen
oxygen *107* Sauerstoff
ozone layer *23* Ozonschicht

P

page someone *123* jdn ausrufen
painter & decorator *17* Maler und Dekorateur
partial(ly) *75* teilweise
participate in *58* teilnehmen an
particle *109* Teilchen
partition *87* Abteilung
pass through *78* durchlaufen, passieren
pattern *17* Muster
paving *17* Fliesen
peak *32* Gipfel, Spitze(nzeit)
pellet *108* Kügelchen, kleine Kugel
perforated *107* durchlöchert
perform *73* ausführen, erledigen, durchführen
(high) performance *28* Hochleistungs-
permanent *90; 103* permanent, Dauer-; ständig, dauerhaft
permission *120* Erlaubnis, Einverständnis, Genehmigung
permit *68* erlauben, zulassen
phenomenon *41* Phänomen
phonetic transcription *76* Lautschrift
pick up the thread *101* den Faden wieder aufnehmen
pier *101* Pier, Pfeiler
piezo electric device *70* piezoelektrische Vorrichtung
pile up *114* (sich) anhäufen
piping arrangement *79* Rohrverbindung
piston *82* Kolben
plant *7* Betrieb, (Industrie-)Anlage, Werk
plastic extruding *72* Plastikpressen
plc *17* (an der Börse notierte) Aktiengesellschaft, AG
plentiful *107* reichlich (vorhanden)
(pair of) pliers *66* (eine) Zange
plug *92* Stöpsel, Stecker
plumbing *16* sanitär(e Anlagen), Installationen
plunge *33* abstürzen, Absturz
poetics *41* Kunst
polar orbit *122* Umlaufbahn über den Pol
polarity *88* Polarität
pole *90* Pol
pollutant *113* Schadstoff
polymer *110* Polymer(e)
portable *123* transportabel, tragbar
position *68* lokalisieren, anordnen
possess *102* besitzen
postpone *119* verschieben
potable *77* trinkbar
potential *50* möglich, potenziell
power socket *121* Steckdose (für Strom)
precast *100* vorfabrizieren
precede *120* vorausgehen, stehen vor
precisely *64* genau
precision *41* Genauigkeit
predetermined *63* vorherbestimmt
prediction *47* Vorhersage
preliminary inquiry *15* Voruntersuchung
preparatory *60* vorbereitend
pressurized-water reactor 84 Druckwasserreaktor
presumably *124* vermutlich
pretend *119* vorgeben, so tun als ob
prevention *107* Verhütung
previous *113* vorhergehend, bisherige/r/s
primarily *37* vor allem, hauptsächlich
prior to *58* vor
prise out *93* heraushebeln
probability *84* Wahrscheinlichkeit
(im)probable *23* (un)wahrscheinlich
productivity *71* Produktivität
profile *11* Profil, Porträt, Beschreibung
prominent *55* prominent, hervorstechend
promotion *55* Beförderung
promotional *11* Werbe-
prompt(ly) *9* unverzüglich, sofort, pünktlich
proper(ly) *17* richtig, ordnungsgemäß
proportion *81* Teil, Anteil
propose *26* vorschlagen
pros and cons *86* Für und Wider
prospect *12* Aussicht
prosperity *115* Prosperität, Wohlstand
protocol *124* Protokoll
proverbial *97* sprichwörtlich
provider *28* Lieferant
proximity *120* Nähe
public sector *37* öffentliche Hand, öffentlicher Sektor
pulp *50* zerkleinertes Erz
pulse *117* Puls
pumped draindown system *79* Ein System, in dem das Wasser bei Frostgefahr abgelassen wird und später wieder mit einer Pumpe aufgefüllt wird.
purchase *11* Kauf, Einkaufs-
purely *41* ausschließlich
purification *107* Reinigung, Säuberung
purify *107* reinigen
purity *115* Reinheit
pursue *58* verfolgen, nachgehen
pylon *100* Mast

Q

qualified by *43* qualifiziert durch, bestimmt durch
quill *97* Spule, Federkiel
quotation *16* (Kosten-)Voranschlag, (Preis-)Angebot

R

radiator *22* Heizkörper, Kühler
railway terminal *35* Endbahnhof
raise *86; 104* erhöhen; aufrichten
raised *36* erhöht, erhaben
random *118* Zufalls-
random(ly) *70* zufällig
range *7* Sortiment, Auswahl

rate *74* Rate, Prozentsatz
ray *117* Strahl
rearrange *26* umstellen, umändern
reassuring *124* vertraut
rebuild *50* wiederherstellen
receiver *117* Empfänger
recent *122* kürzlich, jüngst
reciprocating *68* sich hin und her bewegend
recirculate *85* umleiten
recommendation *22* Empfehlung
recondition *50* überholen, erneuern
reconsider *51* überdenken
re-cover *17* neu furnieren, neu beziehen
recover *33; 114* sich erholen; wiedergewinnen, regenerieren
recovery *33* Erholung, Aufschwung
recreate *124* neu (er)schaffen
rectangular *97* rechteckig
rectangular frame *88* rechteckiger Rahmen, Rechteck
recyclability *110* Wiederverwertbarkeit
recyclable *114* wiederverwertbar, recycelbar
recyclables *108* wiederverwertbare Stoffe, Wertstoffe
recycler *111* Wiederverwerter
recycling *18* Recycling, Wiederverwertung
redesign *73* umstrukturieren
reduce *7* reduzieren, verringern
reduction *12* Verringerung
refer to *43* sich beziehen auf, verweisen auf
reference number *38* Aktenzeichen, Kennziffer
referred to *117* bezeichnet
refill-container *51* Nachfüllpackung
refine *101* raffinieren, verfeinern
refinement *113* Verbesserung
refit *92* wieder einsetzen
reform *110* umformen
reformulate *68* umformulieren
regarding *38* betreffs
regenerator *118* Regenerator
registered with *17* registriert bei, eingetragen bei
regroup *116* umgruppieren
reheat *110* wieder erwärmen
reinforce *98* verstärken
reinforcing bar *17* Verstärkungsstange
reject *125* zurückweisen, abweisen
relating to *53* in Bezug auf, in Zusammenhang mit
relative(ly) *36* relativ, verhältnismäßig
relay *122* Relais, übertragen
release *67* freilassen, freigeben, entlassen
reliability *11* Zuverlässigkeit
reliable *15* zuverlässig
reluctance *28* magnetischer Widerstand
rely on *62* sich stützen auf
remarkable *102* bemerkenswert, beachtlich
remote *122* (weit) entfernt
remove *49* entfernen, beseitigen
renovate *15* renovieren
renovation *15* Renovierung
repeatability *28* Wiederholbarkeit
repeater *118* (Leitungs-)Verstärker
replace *7* ersetzen, austauschen
replaced by *29* ersetzt durch
replacement *16* Ersatz, Austausch, Ersatzteil
represent *33* repräsentieren, darstellen
represent as *118* darstellen als
representative *7* (Handels-)Vertreter/in
reputation *28* Ruf
require *32* benötigen, erfordern
requirement *30* Bedingung, Bedürfnis
resale *50* Wiederverkauf, Weiterverkauf
research *11* Forschung
researcher *114* Forscher/in
reservation *27* Reservierung, Buchung
reshape *110* umgestalten, umformen
residence *36* Wohnanlage
residue *50* Rückstand, Rest
resist *97* widerstehen, sich wehren (gegen)
resource *17* Mittel, Reserve
respective *56* jeweilig
responsibility *42* Verantwortlichkeit, Verantwortung
restore *9* restaurieren, wiederherstellen
restrict *69* einschränken
restricted to *82* beschränkt auf
restructure *12* umstrukturieren
retain *17* zurückhalten, (be)halten
retooling *74* Neuausrüstung (mit Werkzeugen oder Maschinen)
retrieve *118* wiederherstellen, wiedererlangen
reuse *50; 114* Wiederverwendung; wieder verwenden
reveal *81* enthüllen, zeigen
reversal *117* Umkehr(ung)
reverse *88* umkehren
revolution *95* Drehung
revolutionize *62* revolutionieren
rewards package *55* Gesamtheit der Vergütungen
(non) rewirable *93* (nicht) neu anschließbar
rewrite *35* umschreiben
rib *97* Rippe
right angle *90* rechter Winkel
rigid *104* starr
rigidity *98* Stabilität
riser *79* Steigleitung, Steigrohr
rivet *101* Niete
robotic *13* Roboter-
rod *72* Stab, Stange
(felt) roofing *16* Dachdeckerarbeiten (mit Dachpappe)
roughly *74* ungefähr
rub *90* reiben, schleifen
run up *17* hinauffahren, sich erstrecken
running centre *66* Laufzentrum

S

sack *73* feuern, entlassen, rauswerfen
sample *107* Muster, Probe
satisfactorily *85* zufrieden stellend
scale *33* Skala, Maßstab
schedule *26* Zeitplan
scheme *17* Plan, Programm
scientific *110* (natur)wissenschaftlich
scrapyard *51* Schrottplatz
screech *124* Schrillen
screen *107* (Gitter-)Sieb, (Draht-)Gitter
screwdriver *92* Schraubenzieher
screwhole terminal *92* Schraubanschluss
scrubber *109* Teerscheider, Rieselturm
seal *92* Siegel, versiegeln
security bead *17* Sicherheitsrandversteifung
segment *74; 95* unterteilen; Teilstück, Segment
segregate *49* trennen
select *18* auswählen
selective *74* wählerisch
self-enforcing *68* selbsttätig
semicircular *95* halbkreisförmig
semi-detached house *17* Doppelhaushälfte
semi-finished *7* Halbfertig-
sensible *18* vernünftig
sensibly *16* vernünftig, sinnvoll
sensitive *18* empfindlich, sensibel
sentence switchboard *85* Satzbautafel
sequence *104* Reihenfolge
sequencing *52* Reihenfolge, Anordnung
serial number *123* Seriennummer
service center *123* Service Center
servo-mechanism *72* Servomechanismus
sharpen *67* schärfen
sharply *32* rasch, stark
shingle *107* Schotter
shore *100* Ufer
shortage *115* Knappheit, Mangel
shortly before *32* kurz vor
shred *50* shreddern, zerkleinern, zerfetzen
sideways *105* seitwärts, zur Seite
sift *109* (durch)sieben
significant(ly) *73* bedeutend, merklich, erheblich
simplify *97* vereinfachen
simultaneous(ly) *32* gleichzeitig
site *36* (Bau-)Stelle, Platz, Gelände
site on *85* platzieren auf, stationieren auf
slanted *79* geneigt, schräg
sleeve *31* Hülse, Büchse, Muffe, Buchse, Manschette
slenderness *104* Schlankheit
slide along *111* entlanggleiten
slide back *67* zurückgleiten
slide down *94* hinabgleiten
slight *95* gering, leicht
slip ring *90* Schleifrad
slipformed *104* Gießverfahren für Beton
slipping *70* Rutschen
slot *95* Nut(e), Kerbe
slung from *104* herabhängen von, aufgehängt an
smart *123* intelligent, klug
smelt *50* (ein)schmelzen, verhütten
snag *30* Schwierigkeit
snorkel *82* Schnorchel
socket outlet *93* Steckdose
sodium hydroxide *114* Natriumhydroxyd, Ätznatron
soften *33* abschwächen
solar panel *23* Sonnenkollektor
solder *91* löten
solder joint *70* Lötstelle
solely *82* einzig und allein
solid *98* fest(er Körper)
sophisticated *32* hoch entwickelt, raffiniert
sorting technique *108* Sortiertechnik, Scheidetechnik
sour on *74* sauer werden in Bezug auf
spacing *98* Abstand
span *97* umspannen
spark *73* Funke
speciality *16* Spezialität, Spezialgebiet
specialize in *15* sich spezialisieren auf
specifically *56* speziell, bestimmt
specification enclosure *29* beigefügter Bauplan
spectacular *104* sensationell, spektakulär

speculate *51* spekulieren
spin *91* spinnen, (schnell) drehen
spin together *102* zusammendrehen
spokesperson *44* Sprecher/in
spotty *125* ungleichmäßig, uneinheitlich
square meter *81* Quadratmeter
stability *28* Stabilität
stabilize *103* stabilisieren
stable *90* stabil
starter *50* Anlasser
state-of-the-art *45* Stand der Technik
static *97* statisch
stationary *95* fest(stehend), stehend, parkend
steadily *32* stetig, regelmäßig
steady-state *28* Ruheposition
steering wheel *74* Lenkrad, Steuer
stiffen *104* (ver)steifen, (ver)stärken
stillness *107* Stille
stopover *35* Zwischenlandung, Reiseunterbrechung, kurzer Aufenthalt
storage reservoir *107* Speicherbecken
strain gauge *70* Dehnungsmesser
strand *92* Strang, Strähne
strategy *51* Strategie
stray whiskers *92* lose Drähtchen
stress *41* betonen, hervorheben
strictly speaking *111* genau genommen
striking *36* bemerkenswert, eindrucksvoll
strip *92; 108* entfernen, abisolieren; Streifen
strip out *50* demontieren, herausnehmen
stronger presence *56* stärkere Präsenz
structural *97* strukturell, Bau-, Konstruktions-
subassembly *49* Untergruppe
subcontractor *104* Subunternehmer
subject to *101* unterwerfen
subjected to *111* unterworfen, ausgesetzt
subscriber *62* Abonnent/in
subsidiary *29* Tochtergesellschaft
substance *113* Substanz, Stoff
substantial(ly) *29* wesentlich, beträchtlich
substitute *93* Ersatz-
sub-system *118* Untersystem
subtract *73* abziehen, verringern
sub-zero *79* unter Null
succinctly *124* prägnant, in Kürze
sufficiently *113* ausreichend, genügend, genug
sulphur dioxide *113* Schwefeldioxid
sum up *52* zusammenfassen
supercritical *114* superkritisch
superimpose *80* hinzufügen, zur Deckung bringen
superior *17* hochwertig, hervorragend
supersede *101* ersetzen, ablösen
superstructure *101* Aufbau
supplement *107* ergänzen, ersetzen
supplier *7* Anbieter, Lieferant/in
supply *24* beliefern, liefern, versorgen
suppress *124* unterdrücken
surprising(ly) *125* überraschend(erweise)
surroundings *18* Umgebung
survey *17* Gutachten, Untersuchung
suspect *51* vermuten
suspend *68* aufhängen
suspended *101* hängend, schwebend
suspended matter *107* Schwebstoff(e)
suspension bridge *101* Hängebrücke
sustain *84* aufrechterhalten
swaying *104* Schwingen, Schwanken
switchboard *112* Satzbautafel
(forward/reverse) switching *29* Umschalten (von vorwärts auf rückwärts)
syllable *76* Silbe

T

tactile *69* tastbar, Tast-
take for granted *70* als selbstverständlich ansehen, betrachten
tap *107* Wasserhahn
tap into *123* sich einschalten, anzapfen
tear out *15* herausreißen
technician *116* Techniker/in
technological *45* technologisch
telescopic mast *103* Teleskopmast
temporary *101* provisorisch, Hilfs-
tend *74* bedienen
tend to *32* neigen zu
tensile strength *98* Dehnfestigkeit, Zugfestigkeit
tension *98* (An-)Spannung
tensioned *104* Spann-
tensioning *104* Spannung
term *13* Ausdruck, Bezeichnung
territorial subdivision *62* territoriale Unterteilung
textiles *114* Textilien
thermal *17* thermisch, Wärme-
thermoplastic *110* thermoplastisch(er Stoff)
thermosetting *110* durch Wärme härtbar
thoroughly *113* gründlich
thread *67* einfädeln, (hindurch)ziehen
throughout *11* überall (in)
throw away *16* wegwerfen
throw-away society *112* Wegwerfgesellschaft
thus *30* so(mit)
tidal *82* Gezeiten-
tide *82* Gezeiten, Flut
tighten *92* anziehen
tilting pressure *104* Kippdruck
time-consuming *42* zeitraubend
timing *47* Timing, Zeitplanung
to such an extent *70* in solchem Ausmaß
torch *93* Taschenlampe
touchy *53* empfindlich, heikel, kritisch
tough *115* widerstandsfähig, zäh, hart
toughest *113* schärfst, härtest
toxic *82* giftig, toxisch
track *73* verfolgen
tracking *80* verfolgend
transcript *80* Abschrift, Niederschrift
transducer *118* Umsetzer, Umwandler
transfer *62* übermitteln, übertragen, verwandeln
transform *68* verwandeln
transformation *68* Umwandlung
transmission *117* Übertragung, Sendung
transmit *63* übertragen
transmitter *62* Sender
transparent *88* durchsichtig, durchlässig
transportation *115* Beförderung, Transport(wesen)
trash *109* Abfall, Müll
traverse *62* durchqueren
trial *29* Test, Versuch
triangle *118* Dreieck
trim *17* Innenausstattung
truss bridge *101* Fachwerkbrücke
tubing *84* Röhren
turning *66* Drehen, Drechseln
twist together *92* zusammendrehen
two-way data pager *123* Datenüberträger und -empfänger

U

undertake *16* übernehmen
undoubtedly *102* zweifellos, ohne (jeden) Zweifel
unique *28* einzigartig
unnoticeable *63* unmerklich
unobtrusively *107* unaufdringlich, unauffällig
unplug *93* den Stecker herausziehen
unscrew *92* abschrauben
unstable *98* nicht stabil
upside-down *122* verkehrt herum, umgekehrt
uranium dioxide *84* Urandioxid
urban *123* städtisch, Stadt-

V

valve *78* Ventil
vapour *87* Dampf, Dunst
variable *84* veränderlich, regelbar
variation *34* Variation, Abweichung
vary *14* unterschiedlich sein, wechseln, sich unterscheiden, variieren
vent *17* Lüftung
ventilate *28* ventilieren, (be)lüften
venture *124* (sich) wagen
vessel *18* Gefäß, Behälter
via *123* über, durch
vibrating *117* vibrierend
virtually *45* praktisch, nahezu
visual *41* visuell
visual sensing *69* optische Erfassung
vital *104* (lebens)wichtig, wesentlich
voltage *63* Spannung
vulcanisation *114* Vulkanisierung

W

waste incineration *109* Müllverbrennung
wasteful *16* verschwenderisch
water treatment *106* Wasseraufbereitung
wave-like *102* wellenförmig
wear and tear *29* Abnutzung, Verschleiß
wedge *101* Keil
weld *73* schweißen
wherever *11* wo (immer)
wind around *96* wickeln um
winding *91* Windung
wing *36* Flügel
wire drawing *66* Drahtziehen
wireless *122* drahtlos, Funk-
wiring *50* Verdrahtung
wisdom *73* Weisheit
withstand *97* widerstehen, Widerstand leisten
woodgrain *17* Holzmaserung
workforce *12* Belegschaft, Angestellte
workmanship *17* Arbeit(squalität)
workpiece *66* Werkstück
worktable *69* Werktisch
wrought iron *101* Schmiedeeisen, Schweißeisen, Stabeisen

Z

zinc oxide *114* Zinkoxid
zinc sulphide *114* Zinksulfid
zirconium *84* Zirkonium

Mathematical operations

1, 3, 5	are odd numbers.
2, 4, 6	are even numbers.
1, 2, 3	are whole numbers.
$\frac{1}{2}, \frac{1}{4}, \frac{1}{6}$	are fractions.
1.1	one point one
0.1 (.1)	nought (zero) point one
1.047	one point zero (o) four seven
212.5897	two hundred and twelve point five eight nine seven
21.63333333	twenty-one point six three recurring
$a > b$	a is greater (larger) than b.
$a < b$	a is smaller than b.
$2 + 6 = 8$	Two plus six is (equals) 8.
$9 - 3 = 6$	Nine minus three is (equals) six.
$3 \times 2 = 6$	Three multiplied by two *oder* Three times two is (equals) six.
$16 \div 4 = 4$	Sixteen divided by four is (equals) four.
$a \pm b$	a plus or minus b
$\frac{a}{b}$	a divided by b
$2^2 = 4$	Two squared is (equals) four.
$2^3 = 8$	Two cubed is (equals) eight.
$3^4 = 81$	Three to the power of four *oder* Three raised to four is (equals) eighty-one.
$\sqrt{9} = 3$	The square root of nine is (equals) three.
$\sqrt[3]{27} = 3$	The cube root of twenty-seven is (equals) three.
$\sqrt[n]{a}$	The nth root of a
$(3 + 4)^2 = 49$	Three plus four in brackets (parentheses) squared is (equals) forty-nine.
$x'y''$	x dash times y double dash
$x(y + z)$	x open bracket y plus z close bracket *oder* x times the sum of y and z
$\frac{dy}{dx} = \tan \alpha$	The first derivative of y with respect to x equals tangent alpha *oder* dy (divided) by dx equals tangent α.
$\log \sin \alpha$	logarithm of the sine of the angle alpha
$\lim f(x)$	limit of the function of x
$\Sigma_k = \frac{F}{A}$	Stress is defined as force divided by area.
$\frac{F}{A} = \frac{\pi^2 E I}{A l^2}$	F divided by A equals Pi squared multiplied by the module of elasticity times the minimal moment of inertia divided by the cross section times the length squared.

SI/ISO units and conversion table

Quantity (Symbol)	SI / ISO Unit	Internat. symbol	Conversion
Time (t)	day hour minute second	d h min (′) s (″)	
Length (l)	metre nautical mile	m n mile	1 mile = 5280 ft = 1609.344 m 1 n mile = 1852 m 1 yard (yd) = 3 ft = 36 in = 0.9144 m 1 foot (ft) = 12 in = 0.3048 m 1 inch (in) = 25,38 mm
Area (A)	square metre	m^2	1 yd^2 = 0.836 m^2 1 ft^2 = 0.0929 m^2 1 in^2 = 644.14 mm^2 1 cm^2 = 0.155 in^2 144 in^2 = 1 ft^2
Volume (V)	cubic metre	m^3 litre	1 yd^3 = 0,7645 m^3 1 ft^3 = 0.02831 m^3 1 in^3 = 16.34 cm^3 **UK:** 1 gallon (gal) = 277.42 in^3 = 4.546 litres = 8 pints (pts) = 160 fluid ounces (fl oz) 1 quart (qt) = 2 pints = 1.14 litres **US:** 1 gal = 231 in^3 = 3.785 litres (dm^3) = 8 liquid pints (liq pt) = 128 fluid ounces (fl oz) 1 barrel (oil) = 42 gal = 159 litres 1 quart = 2 pints = 0, 946 litres
Velocity (v) Acceleration (a)	metre per second kilometre per hour knot metre per second	m/s km/h kn m/s^2	foot per second (ft/s) = 0.3048 m/s mile per hour (mph) = 0.44704 m/s 1 km/h = 0.277 m/s 1 kn = 1 n mile per hour = 0.5144 m/s foot per second squared = 1ft/s^2 = 0.3048 m/s^2
Mass (m)	kilogram tonne	kg t	1 pound (lb) = 0.454 kg 1 ounce (oz) = $\frac{1}{16}$lb = 28.35 g 1 stone (st) = 14 lb 1 hundredweight (cwt) = 112 lb
Force (F)	newton	N	1 N is that force which, when applied to a body with the mass of 1 kg, gives an acceleration of 1m/s^2
Pressure (P)	pascal bar	Pa bar	1 Pa = 1 N/m^2 1 bar = 10^5 Pa = 14.5 psi 1 pound force per square inch (lbf/in^2 oder psi) = 684.76 Pa = 0.0689 bar

Quantity (Symbol)	SI / ISO Unit	Internat. symbol	Conversion
Work, Energy, Power (W,E,P)	joule watt hour watt	J Wh W	1 Nm = 1 J = 1 Ws 1 Wh = 3.6 x 10^3J 1 Kwh = 3.6 MJ (megajoule) 1 W = 1 J/s = V x A 1 ft lbf/s = 1.3558 W 1 horsepower (hp) = 550 ft lbf/s = 745.7 W 1 PS = 745.7 W
Temperature (t)	degree Celsius	°C	1 degree Fahrenheit (°F) = $\frac{(9t°C)}{5}$ + 32 1 degree Celsius (°C) = (t°F - 32)$\frac{9}{5}$
Electric Current (I) Electromotive Force (E) Resistance (R)	ampere volt ohm	A V Ω	$\Omega = \frac{V}{A}$
Fuel Consumption			**USA:** l/100 km = 235.21 ÷ mpg (miles per gallon) mpg = $\frac{235.21}{l/100km}$ **GB:** l/100 km = 282.481 ÷ mpg mpg = $\frac{282.48}{l/100km}$

Quellennachweis

Fotos:
Titel: Tony Stone Bilderwelten/Nadia Mackenzie
Jim Austin: S. 15, 19
Birmingham Convention & Visitors Bureau: S. 48 (4)
British Tourist Authority: S. 47, 48
Das Fotoarchiv: S. 61/Christoph
Deutsche Lufthansa AG: S. 121
Helga Lade Fotoagentur: S. 15/Steeger, 20/Steeger, 30/Bramaz
Uta Hübner: S. 114
IFA-Bilderteam: S. 7/Int. Stock
Renewable Energy Slide Library: S. 113
The Stock Market: S. 54/Lewine

Grafiken:
A History of Mechanical Engineering, A.F. Burstall: S. 66, 67, 101
Architectural Record 4/95: S. 36, 40
Architectural Record 10/94: S. 41, 42
Auto Engines and Electrical Systems: S. 95
Design and Technology: S. 72, 97, 98
Dictionary of Civil Engineering, John Scott: S. 101, 102
Energy A Guidebook, Janet Ramage. S. 80
Explaining Physics, S. Pople: S. 90, 91, 117
Graseby Controls, Oulton Works: S. 24, 28, 29
Manufacturing Engineering Technology, S. Kalpakjinan: S. 69, 72
Materials and Process in Manufacturing: S. 13
Oxford Designers & Illustrators: S. 11, 24, 28, 31, 45, 47, 49, 52, 53, 59, 62, 63, 66, 77, 78, 79, 82, 92, 93, 100, 103, 109, 116, 123, 125, 129, 132
Perspectives on Energy, L.C. Ruedisli: S. 84
Popular Science 10/90: S. 122
Science to Sixteen, S. Pople: S. 89
Technology Made Simple, D. McCloy: S. 72
Telecommunications, John Allen: S. 118
Thames Water, London: S. 106
The world's great bridges, H. Shirley Smith: S. 102

Texte:
Graseby Controls, Oulton Works: S. 24, 28, 29
New Scientist 29.8.92: S. 113
New Scientist 12.3.94: S. 114
Newsweek 27.6.94: S. 73, 74
Thames Water, London: S. 107